The Jenolan Caves

By

Samuel Cook

Edition : Culturea (culurea.fr), 34 Hérault
Contact : infos@culturea.fr
Impression : BOD, Norderstedt (Allemagne)
ISBN : 9791041984473
Date de publication : juillet 2023
Mise en page et maquettage : https://reedsy.com/
Cet ouvrage a été composé avec la police Bauer Bodoni

THE JENOLAN CAVES.

CHAPTER I.

HOW THE CAVES WERE DISCOVERED.

The Jenolan Caves contain some of the most remarkable and beautiful objects in Australian wonderland. They are formed in a limestone "dyke," surrounded by magnificent scenery, and hide in their dark recesses natural phenomena of rare interest to the geologist, as well as of pleasurable contemplation by non-scientific visitors; while in and about them the moralist may find

"— — tongues in trees, books in the running brooks,Sermons in stones, and good in everything."

To see these caves once is to create a lifelong memory. The pink and the white terraces of New Zealand, which before the recent eruptions attracted so many tourists, did not excel in splendour the caves at Jenolan. But it is common for people to go abroad to admire less interesting things than are to be found within easy distance of their starting point, and which, if they were a thousand miles away, would probably be regarded as worthy of a special pilgrimage. There are persons living two or three leagues from the caves who have never seen them, and who, if they embraced the opportunity for inspection, would possibly regard them with the kind of wonder with which they would gaze upon the transformation scene at a pantomime. And yet the most frequent entry in the visitors' book is that the caves are "grand beyond expectation," and in some of their principal features "indescribably beautiful."

The first of these caves was discovered in 1841 by James Whalan, who lived on the Fish River, near what is now the Tarana Railway Station. Having been robbed by a man named McEwan, he accompanied a police officer in search of the desperado, and tracked him to the romantic spot which forms the centre of the cave reserve, where he was captured.

It is possible that some of these caves were known previously to outlaws, who found in them a secure and convenient hiding-place when hotly pursued. But the visit of Whalan on the occasion of the capture of McEwan first brought them into public notice. The name of the bush-ranger is given to the creek which plays an important part in connection with the caves. One of the principal features received its name from the captor, and another—the Bow Cave—is called after some stolen bullock-bows found therein. They were then known as the "Fish River Caves," or as the "Binda Caves." They were called the Fish River Caves because they were in what was then regarded as the "Fish River District," and not very far from the Fish River post-office. They were called the "Binda Caves" after a station about nine miles distant to the northward. And so they were indifferently known until the month of August, 1884, when their designation was officially changed to "Jenolan"—that being the name given by Sir Thomas Mitchell (Surveyor-General of New South Wales) to the mountain from which is named the parish within which the caves are situated.

The change of nomenclature was recommended on the ground that the then existing names were infelicitous and misleading,—the caves being not upon the Fish River, but upon a different watershed, separated from it by the main dividing range of the colony. The time was considered appropriate for remedying the mistake, because a map of the parish of Jenolan was then in course of preparation, and would shortly be lithographed and issued to the public. The official correspondence on this subject discloses the fact that "Binda" was first thought of as a good official name, and then "Bindo;" but the former was found to be the name of a post town between Yass and Goulburn, and the latter the name of a village and a mountain nine miles north of the caves, and, like the Fish River, on the west side of the dividing range.

Some exception was taken to the proposed change. It was urged that the name "Jenolan" was already applied to a mountain in the Capertee district; but to this it was replied that the mountain called "Jenolan," seven or eight miles from the caves in an easterly direction, was marked on Sir Thomas Mitchell's engraved map of the central portion of New South Wales, whilst the other was not so defined—showing precedence in point of time and importance; and, further, that the orthography of the two names is different, the one being spelt "Jenolan" and the other "Geenowlan"—the former being the name of the parish in which the caves are situated, and the latter the name of a peak near Capertee, in the county of Roxburg. So it was finally determined to change the name of the caves to "Jenolan," and in August, 1884, they were gazetted accordingly. Such a change could not have taken place without inconvenience and some misconception. Recently inquiries have been made as to whether the "Jenolan Caves" are newly-discovered wonders, or old friends under a new designation?

For about a quarter of a century after the discovery by Whalan, little notice was taken of the caves. They were regarded by a few who knew about them as remarkable freaks of nature, but allowed to remain unexplored until some of their hidden beauties were so disclosed as to rouse the enthusiasm of the present curator—Mr. Jeremiah Wilson—to whom, for his daring, energy, and patient investigation the public are greatly indebted. When their fame began to be bruited about, the number of visitors increased, and among them were Goths and Vandals who did not scruple to remove many a crystal gem from the still unfathomed caves. It became evident that unless something were promptly done to secure these newly-found treasures to the public, and protect them from ruthless hands, their magnificence would soon be destroyed, and the people deprived of a possession which should be a source of delight and instruction to succeeding generations, and excite the admiration of tourists from all parts of the world. The Government did the right thing when it prevented the acquisition by private individuals of the caves and a large area of land around them. It would have been better had the dedication to the public been made earlier.

The Gazette notice reserving from conditional purchase land about the caves with a view to their preservation, bears date 2nd October, 1866, and has appended to it the signature of the late J. Bowie Wilson, who was then Secretary for Lands in the Martin Ministry. The area specially protected is six and a quarter square miles in the county of Westmoreland, and near to it are some important forest reserves.

The official correspondence from 1866 to the present time is not very interesting, having reference principally to suggested improvements; it is very bulky, and shows that a large amount of official interest has been taken in the subject; but the money expended and the work accomplished indicate that hitherto Government and Parliament have had but a faint idea of their obligations in regard to the Jenolan Caves.

CHAPTER II.

THE APPROACH TO THE CAVES.

There are several routes to the caves. That commonly chosen is by way of Tarana, a small township 120 miles from Sydney by rail, and 2,560 feet above the level of the sea. The train journey is through interesting country. Leaving behind the new western suburbs with their elegant villas, stately mansions, and well-kept gardens, the traveller arrives at Parramatta with its quaint old church, its fine domain with sturdy English oaks of magnificent growth, its glimpses of river, its old King's School, and its many evidences of change from the old to the new.

From Parramatta (which is but 14 miles from Sydney), to Penrith, there are farms, and dark-leaved orange groves sweet-scented and laden with golden fruit; villages and townships and little homesteads where peace and contentment seem to reign; orchards and cultivated fields with rich brown soil on the hill sides; fine horses, splendid cattle, and cottages with troops of sturdy children. At Penrith, 34 miles from Sydney and 88 feet above sea-level, the country is flat, and the Nepean River which flows in graceful contour is spanned by a magnificent iron bridge supported on four massive piers of solid masonry. The train speeds across the Emu plains which are walled in by the Blue Mountains, so-called on account of the azure haze which covers them as with a bridal veil and is to the everlasting hills what the bloom is to the peach. Scaling the mountain side by a zigzag road, which is one of the "show" works of the colony illustrative of engineering audacity, in the course of a few miles the train climbs to an elevation of 700 feet. At Blaxland's platform, 42 miles from Sydney, the altitude is 766 feet above sea-level.

The name of Blaxland recalls the fact that it was not until 1813 that a route across the Blue Mountains was discovered. Near to the railway line is the track found by Wentworth, Blaxland, and Lawson, over what had theretofore been regarded as an impassable barrier range to the westward of Sydney. On speeds the train, still rising and rising, and revealing a series of views remarkable for grandeur and the sylvan monotony of the gum tree, until at Katoomba platform, 66 miles from Sydney, the elevation is

3,350 feet above sea-level. And so the journey continues past abrupt rocks, gloomy gorges, sparkling waterfalls, rocky glens, bold bluffs, leafy gullies, fairy dells and vernal valleys, until it descends the Great Zigzag into Lithgow, falling about 700 feet in less than five miles! Lithgow is 96 miles from Sydney, and although the Blue Mountains have been passed, the altitude is still about 3,000 feet. Lithgow is a busy place, apparently destined to become a manufacturing centre. There are already in the vicinity numerous coal mines, potteries, and other works. From this point to Tarana the country varies from agricultural and pastoral to rugged scenery.

The distance from Tarana to the caves is 35 miles, which has to be traversed by coach or on horseback. At present the ordinary course is to take coach on the arrival of the train at Tarana and drive to Oberon the same evening. The road is good, the district agricultural, and the scenery agreeable. The course from Oberon to the caves is through a less settled country, and for a considerable distance through unsettled primeval "bush," occupied chiefly by the wallaby, the opossum, the bandicoot, many varieties of the parrot kind that flaunt their gaudy plumage in the sunlight, and the native pheasant or lyre bird (Menura superba) which is a veritable mimic. The country is broken and mountainous and in winter the temperature is low, with cutting winds and severe frosts; in summer the heat is fervent. These silent forces in conjunction with brawling mountain torrents have been large factors in the production of the natural phenomena which are to be found in the valley dammed by the limestone "dyke" in which the caves are formed.

The approach to the Jenolan Caves at the end of the route viâ Tarana is remarkable for its construction and gradient, as well as for the magnificent scenery which surrounds it. All the way from Oberon the land rises, until an altitude of 4,365 feet above sea-level is attained. Then there is a gradual downward grade, until what is called "The Top Camp" is reached, shortly after which the traveller arrives at a part of the range which he traverses by five zigzags, and descends about 500 yards in a total length of road five chains less than three miles! This thoroughfare is made by cutting into the mountain side; and although with a well-appointed conveyance and careful driving it is safe enough, inexperienced travellers feel a sense of relief when the journey is concluded, and they are set down at the Cave House below, which, notwithstanding that it is in the lowest depth of a mountain recess, is still nearly 2,500 feet above sea-level. Any feeling of nervousness, however, is superseded by a sense of the grandeur of the view. If an occasional glance is given at the steep declivity, and a thought occurs as to what would be the consequence of a mishap, the attention is immediately diverted to some new magnificence in the wildly beautiful panorama, the sight of which alone would almost compensate for so long a journey.

On the return trip, when portly gentlemen ascend on foot this very steep zigzag they pause occasionally to contemplate the beauties of nature and estimate the advantages of pedestrian exercise. Perhaps, also, when they have made the same observation two or three times, they begin to think it possible to carry pedestrian exercise to excess, and that a wire tramway would be convenient. Still, at a second or third glance, they get an excellent idea of the course of the limestone ridge, and a better understanding of the operations of Nature in the excavation of the caves and the production of the wonderful formations they contain.

CHAPTER III.

THE EXTERNAL FEATURES OF THE CAVES.

That portion of the limestone dyke in which the caves are found runs six miles north and south; and the Grand Arch and the Devil's Coach House—the two principal "day caves"—are formed right through the mountain, near the centre, in an easterly and westerly direction. On the one side McEwan's Creek flows towards the Devil's Coach House, and on the other side is a natural watercourse leading to the Grand Arch, which is only a few yards distant from it. Bearing in mind how the watercourses converge towards these two central caves, and with what force, in times of heavy rain, the floods scour them, a good understanding may be obtained of the mechanical causes of the enormous excavations which excite amazement as well as admiration. If a visit be paid in winter, when the frost is sharp and the ground is "hoar with rime;" when every bough and every blade of grass is covered with congealed dew and adorned with forms of crystallisation which rival the rarest beauties of the caves; when rocks are split and crumbled by sudden alternations of heat and cold; there will be abundant illustration of the effect of water and light, and the variation of temperature in causing geological transformation. Limestone is not soluble in water without the addition of carbonic acid. An exploration of the caves, however, shows that the mountains are not composed entirely of limestone, but that other substances constitute part of their bulk. The principal causes of the formation of the "day caves," the Grand Arch and the Devil's Coach House, are the mechanical action of water and the variation of temperature. As regards the interior caves, where night reigns supreme, chemical combination has played a more important part. But the effect of water power is everywhere observable in graceful contours, caused by continuous motion, or in stony efflorescence, produced by intermittent humidity or dryness of the atmosphere. The action of the former is the more marked and striking, the latter more elaborate, and microscopically beautiful. The hygrometric condition of the caves is recorded in lovely forms, which lend enchantment to ornate bowers, sparkling grottoes, and fairy cities.

THE GRAND ARCH.

The Grand Arch runs east and west, and is about 150 yards in length, 60 feet high, and 70 feet wide at its western end. The eastern end is 80 feet high, and about 200 feet wide. Its proportions and outline are gloomily impressive, and rather awe-inspiring. It is like the portico to some great castle of Giant Despair. The eastern end is a marvel of natural architecture, and the wonder is how so spacious a roof can remain intact under a weight so enormous. The rugged walls are varied by many peculiar rocky formations. On the northern side is "the Lion," shaped in stone so as to form a fair representation of the monarch of the forest. "The Pulpit" and "the Organ Loft" are suggestive of portions of some grand old cathedral. Adjacent is "the Bacon Cave," where the formations represent "sides," like so many flitches in the shop of a dry salter. The roof is hung with enormous honeycombed masses of limestone, whose sombre shades deepen to blackness in numerous fissures and crannies and cavernous spaces. As seen from the floor the roof appears to be covered with rich bold tracery, engraved by Herculean hands. Near the basement are huge rocky projections, with deep recesses, which for ages have been the retreat of rock wallabies. Near the eastern entrance, lying on the ground, is a gigantic block of limestone, weighing from 1,500 to 2,000 tons, and which at some remote period fell, and tilted half over. This is evident from the stalactite formation which remains on it. Ascending the precipitous masses on the south-eastern side of the eastern entrance over rocks which are, on the upper surface, as smooth as glazed earthenware, a position is attained from which the magnitude of the ornaments of the roof can be estimated. It is then perceived that what, viewed from the floor of the archway, seemed like natural carving in moderately bold relief, are pendant bodies of matter extending downwards 10 to 15 feet, and of enormous bulk. Along the walls of the arch are caves running obliquely into the mountain 10, 15, and 20 feet, and the bottom of which is thick with wallaby "dust." Out of these caves are passages which enable the marsupials to pass from one rocky hall to another until they find a secure refuge in some obscure and sunless sanctuary. The wallaby dust resembles mosquito powder. Perhaps it would be equally efficacious. It is not improbable that the floors of these caves represent a moderate fortune. The explorer sinks over his boot tops in the fine pulverised matter, which, however, is not odoriferous, and is void of offence if a handkerchief be used as a respirator. The presence of this substance, and the oxidisation of its ammonia, probably account for the saltpetre in the crevices of adjacent rocks, although not absolutely necessary to the result, because, in the absence of such accessories, it is an admitted chemical fact that nitrifiable matter is not commonly absent from limestone. In the Mammoth Cave of Kentucky saltpetre manufacture was carried on to a great extent by lixiviation from 1812 to 1814, and during the Civil War a principal factor in the manufacture of gunpowder was obtained from the same source. Up amongst the rocks, midway between the floor and the roof of the eastern entrance to the Grand Arch, in the midst of the wallaby drives, and near to a haunt of the lyre bird, the present curator of the caves had his sleeping-place for 20 years. There he strewed his bed of rushes or of grasses and ferns and mosses; and certainly neither Philip Quarll nor Robinson Crusoe had ever a more magnificent dormitory.

Near to it is a sepulchral-looking place, which, before the Cave House was erected, was reserved for strong-minded lady visitors, and fenced off with a tent-pole and a rug. Farther on is a series of rocks, where bachelors could choose for pillows the softest stones in the arch and dream of angels. All these historic places are pointed out by way of contrast to the state of things now existing, and which, perhaps, in turn will form as great a contrast to the state of things 20 years hence.

From this part of the archway a much better view of the Pulpit, the Lion, and the Organ Loft can be obtained than is possible from the floor. Their massiveness is brought out with great effect. The stalactites and stalagmites which form the organ pipes taper with remarkable grace, and are set off by the shadows in the recesses which vary from twilight grey to the darkness of Erebus. Over all are ponderous masses of blue limestone, with immense convexities filled with perpetual gloom. The rocks leading to the caves, the upper part of which is smooth as glass, owe their polish to their long use by wallabies as a track to and from their favourite haunts. Here and there may be detected in the "dust" on the floor the footprint of the native pheasant. There may also be seen and felt boulders and rugged rocks lying about in strange disorder.

Leaving the Grand Archway by the eastern end, the excursionist descends, through a rocky defile interlaced with foliage, into a dry, stony creek, about which are growing some very rare ferns, as well as some which are common, but nevertheless beautiful, and also some handsome native creeping plants. From this point may be seen the pinnacle which rises over the archway to an altitude of about 500 feet. About 50 yards down this dry creek, and about 20 yards below the junction of the roads from the Grand Archway to the Devil's Coach House, is "The Rising of the Water." Here among the rocks in the bed of the creek the water bursts out of the ground like a sparkling fountain of considerable volume, and "gleams and glides" along a romantic dell "with many a silvery waterbreak." And if it does not "steal by lawns and grassy plots," or yet by "hazel covers," or "move the sweet forget-me-nots that grow for happy lovers," it does here and there "loiter round its cresses." Its banks are so steep that its course cannot be easily followed for any great distance, but, without much difficulty, it may be traced until it flows over a rocky ledge into a deep pool, where there is a wire ladder for the convenience of bathers. Thence it chatters on to the River Cox, whence it enters the Warragamba, which joins the Nepean a few miles above Penrith, and about 50 miles below the Pheasant's Nest. It does not, therefore, enter into the Sydney water supply, but passes through the Hawkesbury to the ocean.

CHAPTER IV.

THE DEVIL'S COACH HOUSE.

The road from the Grand Arch to the Devil's Coach House is devious and uneven, with occasional fissures in the ground indicating the entrance to new subterranean marvels. To the right is the mouth of a cave yet unnamed and unexplored. A little farther on, high up in the rocky wall which connects the two converging mountains of limestone, is the Carlotta Arch, which resembles a Gothic window in the grand ruins of some venerable monastic pile, fretted and scarred by centuries of decay. Almost immediately after passing the line of the Carlotta Arch, the visitor arrives at the Devil's Coach House, which runs nearly north and south. It is an immense cave, whose proportions are better gauged than those of the Grand Arch, because the light flows in, not only from the ends, but also from the roof. At a height of some 200 feet, it has a large orifice in the dome, fringed with stately trees, the fore-shortening of which from their base upwards is very peculiar as seen from the bottom of the cave. On the floor are strewn about rocks of black and grey marble, smoothed and rounded by attrition, and weighing from a few pounds to many hundredweight. In flood-time the storm-waters dash these rocks against each other with tremendous force, and the roaring of the torrent resounds like thunder through the cave. In such wild seasons blocks of stone a ton weight or more are moved a considerable distance. The walls are partly composed of black marble with white veins, and some of the boulders on the floor contain marks of fossil shells.

The most magnificent view of the Devil's Coach House is from the interior of the cave near the northern entrance, from which the rise of the arch appears to be upwards of 200 feet. Its roof is fringed with stalactites, and the outlook is into a wildly romantic gully. Stalactites are suspended from the sides of the entrance, and in several places there are stalagmites covered with projections like petrified sponge, while near to them are formations resembling masses of shells commonly found on rocks by the seaside. Some of these combinations might be examined for hours, and yet leave new and interesting features to be discovered. Small pellucid drops glisten at the ends of the stalactites, illustrating the process of their formation. The large stalactites on the roof and small stalagmites on ledges near the floor of the cave, and vice versâ, afford a practical illustration of the theory that where water flows most freely the stalagmites are largest, and where it flows most reluctantly the stalactite formation is the most magnificent.

From an inspection of these two kinds of cave ornamentation it is seen that, whereas the former are porous and free from central tubes, sometimes running in a straight line and sometimes obliquely, the latter are solid, being formed by lamination and not by accretions of matter conveyed through small interior ducts to external points. This cave may appropriately be called the Marble Hall. Portions of the walls are graced with a "formation" from the limestone rocks above, the stucco having flowed in shapes both grotesque and arabesque. Some of the interstices are filled with stalactites and stalagmites of various colours and proportions. Many stalactites on the roof of the mouth of the cave are said to be from 12 to 15 feet long. All around are entrances to numerous interior spaces adorned with stalactites of the most delicate hues. Some are tinged with various gradations of blue; others are of salmon colour, and delicate fawn. Others again are sober grey, and white shaded with neutral tint. The rocks are decorated with little patches of moss, from rich old gold to living green. The harmony of colour is marvellous, and the combined effect unique. Nature herself has so painted and ornamented the cave as to give a lesson to professors of decorative art. The vision of rocky beauty grows upon the imagination the observer until at last it seems like a new revelation of the enchanting effects which can be produced by natural combination.

To the artist this cave presents attractions of a kind not to be found in any other of the wonderful caves of Jenolan, although commonly it receives small attention from visitors, who recognise its grand proportions, but are impatient to witness the more elaborate and brilliant features in the hidden recesses of the mountain.

Why this spacious cavern should be called the Devil's Coach House (except on the lucus a non lucendo principle) few would divine. The name of his

Satanic majesty is often associated with horses and horse-racing, but not generally with coaches and coach-houses. In this connection, however, it is necessary to observe class distinctions. The cavern is not sufficiently monstrous to be used by Milton's personification of the rebellious archangel, nor sufficiently hideous for Burns's "Auld Clootie," with hoofs and horns. Coleridge's devil or Southey's devil (as illustrated in "The Devil's Thoughts" of the one, and "The Devil's Walk" of the other) was neither too grand nor too ignoble to notice coach-houses. But then, he was a sarcastic fiend, for when he "saw an Apothecary on a white horse ride by on his vocations," he "thought of his old friend Death in the Revelations"—which was rather severe on the pharmacist. But leaving the man of drugs—

"He saw a cottage with a double coach-house,A cottage of gentility;And the Devil did grin, for his darling sinIs pride that apes humility."

The cottage at the caves is not particularly "genteel" in appearance. The coach-house is large enough to hold almost as many horses as were kept by Solomon, and as many chariots as were possessed by Pharaoh, and at one end it is "double;" but there was no thought of Pharaoh, or Solomon, or Coleridge, or Southey when it was named.

It was not because this huge place was considered big enough to be the Devil's Coach House that it was called after the Devil, nor because it was thought to be a suitable place for Satan to "coach" his disciples in; nor had the person who named it any intention of paying a compliment to poetic genius. It was called the Devil's Coach House for reasons similar to those which created the nomenclature of the numerous Devil's Pinches and Devil's Peaks, Devil's Mills and Devil's Punchbowls, in various parts of the world. Captain Cook more than a century ago gave the name of the Devil's Basin to a harbour in Christmas Sound, on the south side of Tierra del Fuego, because of its gloomy appearance—it being surrounded by "savage rocks," which deprive it of the rays of the sun. For similar reasons, perhaps, the name of the Devil's Coach House was given to this interesting portion of the Jenolan Caves, which are surrounded by mountains and "savage rocks," and from which the rays of the sun are excluded, except during a few hours per day. In winter the sunshine does not glint on to the roof of the cave house till about 10, and at about 2 in the afternoon the valley is wrapped in shade.

It is 45 or 46 years since James Whalan came suddenly upon the mouth of this cave, and it so impressed him with its rugged grandeur and weirdness that when he returned home he reported that he had been to the end of the world, and had got into the Devil's Coach House. So by that term it is still called, although it has been since named the Easter Cave, because of a visit by some distinguished member of the Government service during Easter, which in New South Wales is now as favourite a holiday time as it was when kept as a festival in honour of the Goddess of Light and Spring. For a short period in the afternoon one end of the cave is flooded with the warm beams of the sun. Then it is at its best, and, as the enamoured hand of fancy gleans "the treasured pictures of a thousand scenes," so, after the bright rays have disappeared, and the cave is seen in the shade or by "the pale moonlight," its beauties change from hour to hour, like shadows on the mountains or the cloud glories of an autumn sunset.

CHAPTER V.

THE NIGHT CAVES.

The "day" caves of Jenolan, although grandly picturesque, are but slightly typical of the interior caverns. As the few bars of harmony dashed off by way of prelude to an intricate musical composition prepare the ear for the movement which is to follow, so an inspection of the external caves trains the faculty of observation for the ready appreciation of the more elaborately beautiful objects in the dark recesses of the enchanted mountain. The contrast between the two is as marked as the difference between the costly pearls of a regal diadem and the rough exterior of the shells which first concealed them.

To explore the "night" caves it is necessary to be furnished with artificial light, and each excursionist is provided with a candle fitted into a holder, the handle of which is like the barrel of a carriage lamp, and immediately underneath the flame is a saucer-shaped guard with the edge turned inwards, so as to catch the drips from the sperm. By means of this arrangement the lights may be presented at almost any angle without doing injury to the caves, except in regard to the smoke, which although slight, is nevertheless in degree perilous to the wondrous purity of the formations. The principal features of the "night" caves are illuminated by the magnesium light, which is rich in chemical rays and burns with great brilliancy. There should not be allowed in the caves any colorific or other light which would cause smoke by imperfect combustion, or emit volatile substances likely to change the interior hues. If their pristine beauty—or as much of it as remains—is to be preserved, the caves ought to be illuminated by electricity, which will neither affect the temperature nor soil the most delicate of Nature's handiwork. The smoke of candles in a quiescent atmosphere like that of the caves, cannot fail, in process of time, to have a deleterious effect. Years ago, in the Wingecarribee country, there were some fairy gullies. The sides were flanked with sassafras columns, the roofs were covered with branches interlaced by creepers that excluded the sunbeams, and on the banks of the creeks which trickled through the centre were tree-ferns of marvellous beauty. Their perfect fronds were lovely—

their growth prodigious; but that in which their charm chiefly lay was their unexampled delicacy of colour. When the natural shade was removed, and they were subject to wind, and rain, and dust, they became commonplace. They grew like other tree-ferns, and were ranked with ordinary things. So with the caves. Their wondrous beauty and attractiveness are found in their freedom from defilement. In their illumination there should be neither smoke nor heat, and it is a question whether within their precincts incense ought to be burned, even to King Nicotine.

CHAPTER VI.

THE NETTLE CAVE.

The Nettle Cave is for the most part a place of twilight. If visitors are incautious in approaching it they will soon come to the conclusion that it has been properly named, for all around are fine clumps of herbaceous weeds with sharp tubular hairs upon vesicles filled with irritating fluid. The sting of a nettle and the sting of an adder resemble each other, but are yet dissimilar. The adder strikes his tubular fang into his prey, but the nettle victim impinges upon the tubular hair which communicates with the acrid vesicle.

The Nettle Cave is reached by climbing 170 feet to the left of The Grand Arch, and if in the ascent the visitor be invited to smell a plant with alternate leaves and racemes of not very conspicuous flowers, it would be well for him to decline with thanks. There are some rough cut steps leading to this cave, and on one side is a galvanised wire rope supported by iron stanchions let into the rocks, which makes the ascent tolerably safe. The road runs between two bluff rocks, which for a considerable distance rise almost perpendicularly, and then curve so as to form a segment of a circle some 150 feet overhead. The cave is barred from wall to wall by a light iron gate sufficient to prevent improper intrusion, not ponderous enough for a penal establishment, but sufficiently pronounced to suggest Richard Lovelace's lines—

"Stone walls do not a prison make,Nor iron bars a cage;Minds innocent and quiet takeThat for an hermitage."

Descending some of the rough stones and winding along a footpath, the tourist descends into a chamber below a magnificent series of rocks covered with beautiful "formation" from the dripping roofs above. This is called "The Willows," because of the resemblance it bears to the graceful and beautiful appearance of the Salix babylonica, on which in the olden time captive Israelites hung their harps and "wept when they remembered Zion." The entrance to this cave is circuitous. First there are some rocky steps to be climbed, and then the road winds through avenues of "willow" formation up to the summit. From this point about 60 or 70 feet down is a funnel-shaped declivity resembling the mouth of an extinct volcano. In some respects it is like the "Blow Hole" at Kiama (a natural fountain, inland, fed by ocean waves which force their way through a water-worn tunnel). Undoubtedly that also is one of the wonders of the world, but some time since it was utilised by the local corporation as a receptacle for dead horses and defunct cattle! From this declivity in the Nettle Cave the visitor naturally shrinks, being dubious as to where his remains would be found if he were to make an uncertain step. In his timorous progress, however, his attention is soon arrested by some splendid stalagmites to the left of the hellish-looking vacuity. One of the most noble is about four feet in diameter at the base, and from 12 to 14 feet high, covered with curiously-shaped ornamentation, and having minute stalactites projecting from the sides. All about it are nodules of delicate fretwork, as lovely as the coral of the ancient sea out of which this mountain was made millions of years ago. On the apex is a gracefully-tapered cone; and hard by is a small stalagmite covered with prickles as sharply defined as those of the echinus. All around are limestone pictures of surpassing loveliness. There is not much variety of colour, but the formation is infinite in its variety. It is intended to have the hideous and perilous-looking volcanic funnel previously mentioned guarded by wirework, which is necessary to ensure the complete safety of sightseers. If an unfortunate wight were to trip, he might fall a distance of about 70 feet, and be shot without ceremony into the Devil's Coach House. One remarkable stalagmite in the vicinity of this infernal shaft is shaped like a hat, and another is like a gigantic mushroom. The floor of the cave is thin, and when stamped upon vibrates in imitation of an earthquake wave. Stalactites in rich profusion depend from the roof, and here and there are clumps of bats, clinging together like little swarms of bees. The stalactites are tipped with drops of lime-water clear as crystal at the lowest point, and becoming gradually opaque. It is also noticeable that while the drops at the ends of the stalactites appear to be perfectly still globular bodies, their molecules seem to be in perpetual motion. The opaque part of the drops thickens until it resembles sperm, and then the gradation is almost imperceptible until it unites with the solid formation. All around are curiously-shaped drives, one of which has been explored until it communicates with the Imperial Cave. It is not an inviting entry, for it is low and narrow, and has sharp stalactites on the roof. The floor is covered with very fine dust, about the eighth of an inch thick, which, however, seems not to rise, and when struck with a hammer the sound is like a blow struck upon a carpet, and the dull thud reverberates in the caverns below.

From the end of the cave, looking towards the mouth, the appearance is particularly wild. The stalagmites in front resemble prisoners in some castle keep, and the part of the cave farther on, upon which the light falls, near to the barred entrance, makes the interior shade seem more gloomy. There is one remarkable pillar about 10 feet in diameter from the floor to the roof of the cave; and seeing that it is about 30 feet in height, and has been made by the constant dripping of lime-water, visitors may speculate as to its age, and statisticians may estimate the number of drips required for its creation. Along the sides of the cave are beautiful pillars. Some are like trunks of trees, gnarled and knotted, and some like elaborately-carved columns. There are grottoes and alcoves, and terraces formed by runs of water; Gothic arches and Etruscan columns, carvings of most cunning elaboration, and stalactites more noticeable for their massiveness than for their grace. There are narrow chasms descending into blackness, through which future discoveries may be made. On the water-formed terraces are numerous stalagmites resembling congewoi and other zoophytes. It seems as though Nature had fashioned the cave after a kaleidoscopic view of the most remarkable objects in marine and vegetable life. At the end of this section the roof rises, and is pierced by an inverted pinnacle. The walls are composed of masses of stalactite formation, imperfectly developed by reason of pressure. Near at hand liquid substances have fallen, and petrified so rapidly as to resemble streaks of lava which had suddenly cooled and formed cords and ligaments like grand muscles and tendons.

The eastern end of the cave runs into the Devil's Coach House, about 120 feet above the coach-house floor. The opening is very beautiful, being ornamented with columns and pinnacles, and the view from this point to the interior of the cave is unexampled. Scores of breaches in the roof and sides can be seen leading to other marvellous places—there being cave upon cave and innumerable changes of formation upon the ground. In rocky basins the débris is largely composed of minute bones. The "remains" may be taken up by handfuls. The teeth of bats and native cats—the vertebræ of marsupials and snakes—the wing-bones of birds, and other fragments of the animal world are mixed together in a mammoth charnel-house, whose grandeur could hardly be surpassed by the most costly and artistically designed mausoleum.

The Ball Room—an upper storey of the Nettle Cave—is reached by mounting twenty-nine steps cut into the rock. Near the eastern entrance are two stalactitic figures fashioned like vultures about to engage in combat. All around the little plateau of Terpsichore are huge stalagmites, resembling domes, crowded together and pressing into one another. Some are set off with stalactites; others are honeycombed. Thence the direction is still upwards, and the ascent is made by means of about 50 wooden steps, with a guard rail on each side. The formations are striking and graceful. Pointing upward is a gauntleted hand and forearm of a warrior of the olden time. There are representations of bewigged legal luminaries and bearded sages like Old Father Christmas or Santa Claus. Some of the columns which support the archway have tier upon tier of stalactites, drooping so as to counterfeit water flowing from a fountain, alternating with stalactite formation like boughs of weeping willow. One prominent stalagmite is like the back of a newly-shorn sheep, with shear-marks in the wool. On the western side is a figure like that of an orator in the act of exhortation. The forehead is bald, long white locks are flowing on to the shoulders, one arm is upraised, and the pose gives an idea of earnestness and force. In front, just below the bust, is a reading desk of stone, the outer edge of which is fringed with stalactites. From this place are steps leading to the arch. They are safe and convenient. Underneath them is still to be seen the wire ladder formerly used to pass from the Nettle Cave to the Arch Cave, and it is easy to understand the trepidation of nervous visitors when they were swaying about on it in mid-air over the dark abyss below. After resting for a moment in the midst of a stalagmitic grotto, the visitor ascends some stone steps towards the Grand Arch, proceeds through a beautiful cavern with Norman and Doric pillars, composed almost entirely of stalagmites, and enters the Arch Caves, which were so called because at that time they were accessible only through the Carlotta Arch. They are now, as previously described, approached through the Nettle Cave by means of the wooden staircase, which was built about three years ago.

CHAPTER VII.

THE ARCH CAVE.

The Arch Cave runs north-westerly from the line of road to the Carlotta Arch, and has a gradual descent. It is about a hundred yards long, and in some places about half a chain wide. The roof is decked with beauty; the floor is covered with dust. There is now but one complete column in the centre, and that is formed by a stalactite which extends in a straight line from the roof to the floor. It is surrounded by a number of other magnificent pendants of a similar kind, more or less ornate, and crowded together in rich profusion. Some of them have grown until they nearly touch bold rocks which jut out from the walls, and the spaces between the larger cylindrical forms are filled by stalactites of various lesser lengths, some of which are figured so as to represent festoons of flowers. The complete pillar tapers from the upper to the lower end. For about two-thirds of the way down it is compounded of several stalactitic lines; the remainder is a simple shaft with irregular surface. To the right of it is a marvellous piece of formation like the head of a lion with the forelegs and the hoofs of a bull, posed so as to resemble Assyrian sculpture.

At one time there were in this cave five pillars as perfect as the one which remains, but in 1860 they were destroyed by a Goth from Bathurst. There are numerous columns of dimensions not so great along the sides of the cave, and at every step appear fresh objects of admiration. Some of the stalactites are resonant, and so is the floor, which, on the thinnest portion, responds imitatively to the tramping of feet. In a passage on the right hand side is a stalactite which the cave-keeper has carefully watched for 18 years, in order to form some idea as to the rate of stalactitic growth. He has always found a drop of water clinging to the lowest surface as though it were ready to fall, and yet during the whole term of 18 years the actual addition to the solid stalactite has been only half-an-inch in length, of a thickness equal to that of an ordinary cedar-covered lead pencil. It is evident, from observation of other portions of this cave, that some formations have been created in a manner less slow. Still, it is probable there are stalactites the growth of which has been more gradual than the one subjected to special scrutiny.

The entrance furnishes an illustration of damage done by careless visitors years ago, and of the necessity for constant care to preserve the caves from destruction. When this cavern was first opened to the public the floor was white as snow. It is now black and greasy, as well as dusty. The change has been brought about by the pattering of feet encased in soiled boots, and by drips from candles and torches used before the present lighting arrangements were adopted. Some of the stalactites have their lower portions damaged in a similar way. But, worse still, an elaborate and very attractive specimen, resembling cockscomb, has been damaged by fracture, and made incomplete by unauthorised appropriation. The porosity of some of the rocks can here be readily distinguished. Their surface is like that of pumice stone. In dry weather the walls are sparkling; in wet seasons they are moist and dull. At the far end of the cave the floor is covered with little indurated lumps with carved surfaces. They are all similar in shape, and vary in bulk from the size of quandong seeds or nuts, of which bracelets are sometimes made, to that of a mandarin orange. Perhaps they were fabricated on the roof and became detached. It is hardly possible they could have been formed where they lie without being joined together in a solid mass.

Here perfect silence reigns. It is so profound as to be almost painful, and the darkness is so dense that when the candles are extinguished the visitor can pass a solid object before his eyes without the shadow of a shade being perceptible. It is not suggestive of the darkness which—

"Falls from the wings of NightLike a feather that is wafted downwardsFrom an eagle in its flight."

Nor yet of "the trailing garments of the Night" sweeping "through her marble halls." There is nothing to give the idea of action. Solitary confinement for 24 hours in such a "separate cell" would drive some men mad. At the end of the cave is a mass of stalactites, through which is a passage leading to "The Belfry," where are some large stalactites, three of which, when struck with a hard substance, sound like church bells. One of them has a deep tone, equivalent to C natural. The others do not vibrate so as to produce perfect notes according to musical scale, nor are their sounds either rich or full.

CHAPTER VIII.

THE CARLOTTA ARCH.

On returning to the mouth of the Arch Cave, the tourist proceeds towards the Carlotta Arch—so named in honour of a daughter of the Surveyor-General of New South Wales, Mr. P. F. Adams, who visited the caves 10 or 12 years ago, and has always taken an interest in their exploration and preservation. Ascending some stone steps, guarded by galvanised wire, an excellent view of the Ball Room to the eastward is obtained. The steps make access easy. Previously the rocks were slippery, in consequence of the polish given to them by the feet of marsupials, and the return journey was accomplished by holding on to a rope, and sliding down the glassy surface.

The entrance to the Carlotta Arch is protected by a wire railing, about 35 feet by 8 or 10 feet. Passing through the iron gate, the visitor finds himself on a little platform. Hundreds of feet below is a gully, rippling at the bottom of which is a rill of water, which sings as it goes, and whose melody, softened by distance, is pleasant as the hum of bees or the carolling of birds. Above is a hoary rock, rugged and bare, with the exception of some clumps of lilies which flourish and bloom in its inaccessible clefts. From this point the tourist ascends the Nettle Rocks for about 60 or 70 feet. Some steps are to be cut here, and certainly they are much needed, for at present the journey is very toilsome and difficult. From the end of the protected portion to the summit—70 or 80 feet—the acclivity may be comfortably surmounted by ordinarily active people. The Carlotta Arch is about 100 feet high and about 70 feet wide, with an interiorfringe of stalactites. The picture seen through it is exceedingly grand, including majestic trees and romantic gullies, huge mountains and immense rocks, with bold escarpments. The walls of the arch are pierced like a fortress. Its entire superstructure represents the union of two mountains by a natural bridge, clad with trees and shrubs and creeping plants which trail gracefully down its sides. On the summit are eucalypti, and conspicuous amongst them is an iron guard for the protection of passengers going over the viaduct. About half-way up this track from the arch to the bridge (which is the concluding portion of the new road from Mount Victoria), a good view can be obtained of McEwan's Creek, where the water has broken through the hills, leaving the limestone rocks and caves sometimes on the one side and sometimes on the other for a distance of three miles up the valley running northerly. To the westward is the Zigzag, leading to the cave-house by the route from Tarana, and from which can be obtained the grandest view en route from Oberon to Jenolan. Here, after having spent an hour or two in the caves, it is pleasant to bask in the golden sunshine and watch the gaudy parrots flit by. From this point to the northward the limestone is visible to its full extent till it is overlapped by higher mountains. It is about three miles in length, by a maximum of half a mile in width. Immediately to the south the limestone dyke is covered, but it crops up again about seven miles distant, and continues on the surface for 15 to 20 miles, in the direction of Goulburn. Near to what is called the Gallery (the approach to the bridge over the arch) is an old gum tree, growing right over the centre of the Devil's Coach House, and 500 feet above the gullies, which can be seen by looking over the precipice. If it were a blue gum tree, "and nothing more," it would be as uninteresting as the "yellow primrose by a river's brim" was to Peter Bell. As a specimen of its kind this tree is a failure; but it happens to be in the centre of the cave reserve, and the "blaze" on it bears the mark, "F 69." From this point the reserve extends two and a half miles east and west by five miles north and south, and is certainly one of the most wonderful areas dedicated to the public.

In the rocks near to the Centre Tree is an orifice called "The Devil's Hole." It pierces the mountain obliquely, but without much deviation from a straight line, and a stone thrown down it takes, according to its weight, from nine to twelve seconds to find a resting-place on the floor of the Devil's Coach House! This is an experiment which should not be tried without precaution, and then only under official sanction, otherwise some serious accident may occur. It would be well to erect notice-boards at this and several adjacent places, warning persons not to cast stones into the caverns, for the whole mountain is full of holes and caves and drives. A piece of rock cast heedlessly into a crevice or perforation in one cave might mean death to a tourist in lower cavernous regions, and there is neither medical man nor coroner within convenient distance. From the bridge (which is guarded by wire ropes) on the western side, the visitor looks down on the Elder Cave; the Zigzag is in front, and below is the sylvan valley from which the "ermin'd frost" has been thawed, and which now "laughs back the sun." To the eastward are in view of the spectator the Nettle and Arch Caves gate, the south entrance to the Devil's Coach House, and the waterfall to the Cave River. In the distance can be discerned a place known as Oaky Camp, or McEwan's Camp, which is of interest in connection with bushranging episodes of the olden time. From the highest point of the hill over the Grand Archway the cave-house can be seen nestling in the valley 500 feet below. Perched upon this pinnacle, with terrible depths on each side and awe-inspiring grandeur at every turn, the beholder is apt to realise how very small is the space he fills in the economy of Nature, how inadequate is language to express deep emotions of the mind, and how marvellous are the works of the Creator!

CHAPTER IX.

THE ELDER CAVE.

On leaving-the Carlotta Arch and the bridge, the visitor—mentally gratified, physically tired, and conscious that his perceptive faculties have been somewhat strained—rejoices that the cave-house is conveniently near, so that he can promptly ensconce himself in an easy-chair and meditate upon the charming scenes upon which his eyes have feasted. If he be unusually robust he may economise the return journey by taking a peep at the Elder Cave, which lies just off his homeward course. It derives its name from the elder trees which grow about it and conceal its entrance, which is at the bottom of a "ragged" shaft similar to that described in the tragedy of "Titus Andronicus," whose authorship is disputed, but which Samuel Phelps and others have no doubt was written by Shakespeare. It resembles the "subtle hole" where Bassianus lay imbrued "all of a heap like a slaughtered lamb." But that was near an alder, and not an elder, tree; and, so far as is known, the pit which leads to the Elder Cave has no associations so tragic as those which are inseparable from the horrible brutalities of "Titus Andronicus." Its mouth is not covered with "rude growing briars," nor are there upon the leaves "drips of new-shed blood as fresh as morning dew distilled on flowers." On the contrary, it is a rather cheerful-looking pit, filled up with foliage like an arborescent bouquet in an enormous natural vase. For a long time its cavity was completely obscured by the leafy covering, and it was first entered by climbing along a branch level with the surface of the ground, and descending the trunk of the tree to the bottom of the well. There are several elders in the pit, which, being unusually moist, is favourable to their growth, and they bear splendid cymes of cream-coloured flowers and black berries suggestive of spiced home-made wine.

The Elder Cave was found by Mr. Wilson in 1856, but it has not had much attention bestowed upon it, probably because its beauties have been eclipsed by later discoveries. The first part consists of some rather large chambers connected by small passages, rough inside, and difficult to explore. All are pretty, and one, named "The Chapel," contains stalactites called "shawls," on account of their resemblance to ladies' vestments so designated. One of these is about five feet long by six inches deep, and a quarter of an inch thick. Half of it is of glassy clearness. The floor is of ornate formation. The next chamber is called the "Coral Cave." It is difficult of access. The way for about 100 yards varies from only two to four feet from floor to ceiling. Nearly at the end is a hole about 12 feet in diameter and 15 feet deep, containing fossil bones. From floor to roof the formation is grand. There are a few fine stalagmites, but the chief beauty is in the stalactitic growth. Many of the stalactites hang from the lowest shelving rock to the floor, and form an alabaster palisade. Immense bunches of snow-white limestone droop from the roof, and one unusually large conical mass tapers off until it connects with the apex of a pyramidal block on the floor. In contrast with these ponderous specimens are numerous straw-like glassy tubes. Portions of the floor are covered with beautiful coral.

Near the mouth of this Pit Cave is an aperture of special interest, because it is the entrance to the shaft at the bottom of which, on the 16th February 1879, the intrepid curator discovered the Imperial Cave, which is one of the most magnificent opened to the public. He made three separate attempts before he was able to bottom this deep black hole. On the first occasion he was lowered into it at the end of a rope, and when all the line had been paid out was dangling in mid-air at the end of his tether. When he was let down a second time with a longer cord it was found to be deficient, and the cave-keeper was still suspended in ebon space. The second failure made him still more resolute. He did not believe that the black hole into which he had descended was the bottomless pit, and so he tried again to fathom its inky depths, and at a distance of 90 feet from the surface alighted upon the rocky floor of what is now called the Imperial Cave.

Cave exploration is not what would be commonly regarded as a pleasant pastime. It requires a lissom body, plenty of physical strength, and a strong nerve to worm along narrow passages, without any certainty of being able to reach a turning-place, and with the risk of being so wedged in as to make retreat impossible. A stout heart is necessary to enable a man to descend to unknown depths of blackness from mouths of fearsome pits, close proximity to which makes one's flesh creep. A fracture of the rope or the falling of a piece of rock might give the explorer his quietus. A somewhat sensational illustration of this kind of peril is given in Griffin's "Studies in Literature." The eldest son of George D. Prentice, one of the sweet singers of the New World, determined to fathom the maelstrom of the Mammoth Cave in Kentucky. A long rope of great strength was procured, and with a heavy fragment of rock attached to it, like a stone at the end of a kellick rope, it was let down and swung about to clear the course of loose stones. "Then the young hero of the occasion, with several hats drawn over his head to protect it as far as possible against any masses falling from above, and with a light in his hand and the rope fastened around his body, took his place over the awful pit, and directed the half-dozen men, who held the end of the rope, to let him down into the Cimmerian gloom. Occasionally masses of earth and rock whizzed past, but none struck him. On his way, at a distance of 100 feet, the spray caused by a cataract which rushed from the side down the abyss nearly extinguished his light. One hundred and ninety feet down he stood on the bottom of the pit. Returning to the mouth of the cave the pull was an exceedingly severe one, and the rope, being ill-adjusted around his body, gave him the most excruciating pain. But soon his pain was forgotten in a new and dreadful peril. When he was 90 feet from the mouth of the pit and 100 from the bottom, swaying and swinging in mid air, he heard rapid and excited words of horror and alarm above, and soon learned that the rope by which he was upheld had taken fire from the friction of the timber over which it passed. Several moments of awful suspense to those above, and still more awful to him below, ensued. To them and to him a fatal and instant catastrophe seemed inevitable. But the fire was extinguished with a bottle of water belonging to himself, and then the party above, though almost exhausted by their labour, succeeded in drawing him to the top. He was as calm and self-possessed as upon his entrance into the pit; but all of his companions, overcome by fatigue, sank down upon the ground, and his friend, Professor Wright, from over-exertion and excitement, fainted, and remained for some time insensible. The young adventurer left his name carved in the depths of the maelstrom—the name of the first and only person that ever gazed upon its mysteries."

The keeper of the Jenolan Caves has had many experiences quite as thrilling as that of the son of George D. Prentice, who, some time after his descent into the maelstrom, fell in the conflict between the Northern and the Southern States of the American Union. The curator has hundreds of times wormed his way in the darkness through narrow drives and descended black holes of unknown dimensions by means of ropes and ladders. He has burrowed about like a rabbit, squeezing through small apertures, occasionally having his clothes torn off him by stalactites, and his knees wounded by miniature stalagmites, and his sides abrased by the sharp corners of projecting rocks. When being lowered by ropes he has run the risk of being brained by falling débris. Fortunately, he has been preserved from serious injury, and is still as lithe as a ferret. Christopher Columbus made wonderful maritime discoveries in the Western hemisphere, and Captain Cook distinguished himself in the Southern seas, but neither the bold Genoese nor the stout-hearted Yorkshireman who thrice circumnavigated the globe could have thrown more earnestness into his work than has been displayed by the subterranean explorer at Jenolan, of whom it may be said, without prejudice to his good name, that he has done more underground engineering than any "road-and-bridge" member of the Legislative Assembly, performed more turning and twisting than the most slippery Minister of the Crown who has ever held a portfolio in New South Wales, and found secluded chambers enough to permit every political or social Adullamite—"every one that is in distress, and every one that is in debt, and every one that is discontented"—to have a little cave of his own. As the visitor has to be guided by the curator through labyrinthine passages as intricate as the most puzzling mazes of Crete or Egypt, in order to see fairy grottoes, crystal cities, jewel caskets, coral caves, and mystic chambers which he has discovered, it may be here recorded that Mr. Jeremiah Wilson was born in Ireland, near Enniskillen, that he was three years old when he came to New South Wales, 43 years ago, and that his family have lived continuously near Oberon. His first visit to Jenolan was with a party of excursionists. He has ever since taken a romantic interest in the caves, and from the time of his appointment as cave-keeper in 1867 until now his enthusiasm for exploratory work appears to have never flagged.

CHAPTER X.

THE LUCAS CAVE.

The Lucas Cave presents, in grand combination, almost every type of subterranean beauty to be found in the natural limestone caves of Jenolan. It rivals the Imperial Cave, which, however, is commonly regarded as the more attractive, and displays a more dazzling magnificence than that which characterises either the Arch or the Elder Cave. The approach to the Lucas Cave is by a zigzag path from the valley, leaving the high Pinnacle Rock to the left hand. The route is not difficult to agile people, but the road would be greatly improved by the cutting of suitable steps. On gaining the top of the ridge the waterfall is in front. To the left are rocks rising like a vast citadel to a height of 900 feet, at the summit of which are immense cliffs with deep gorges between them. The distance is too great to enable the visitor to discern their geological composition. Some of them seem as though they had been shaped by human hands in the time of the Pharaohs. They remind one of the enormous stones in the Great Pyramid of Egypt, or the massive blocks in the Temple of the Sun at Heliopolis, and the limestone ridges suggest the mighty Nile which runs through similar ranges. These elevated pinnacles and chasms are favourite resorts of marsupials. Wallabies may be seen leaping from rock to rock and peering out from the crevices. As they are not molested they afford visitors ample opportunity to watch their graceful movements. The distance from the top of the ridge to the mouth of the cave is about 100 yards, with a fall of 60 feet. The descent in some places is so steep as to make it difficult in dry seasons. In wet weather it is dangerous, the rocks being covered with slippery clay.

The grand cavern, called the "Lucas Cave," was so named in recognition of valuable services rendered by the Hon. John Lucas, M.L.C., who, from the 8th December 1864, until the dissolution of Parliament on the 15th December 1869, represented in the Legislative Assembly the electorate of Hartley, in which Jenolan is situated. He used his influence to obtain the dedication of the reserve, and make provision for the care and improvement of the caves. It was on his recommendation that the present cave-keeper was appointed to the office of curator. His foresight and activity are suitably commemorated by the association of his name with objects of beauty, the fame of which is now spread throughout the whole civilised world.

The opening to the Lucas Cave is very massive, and has a rather steep fall of about 12 feet from the pathway to the floor of the cavern. The entrance is about 30 feet wide and 25 feet high. The roof of the portico is ornamented by rocks, which in shape and colour appear to be in keeping with the gloomy-looking tunnel beyond. The overhanging masses are honeycombed and convoluted in a remarkable manner, and thin off to points like stalactites. The curved, tapering forms are in groups of various dimensions, drooping in folds like those of loosely-fitting garments. They represent not "formation," but the original rock, out of the crevices of which the softer portions and earthy substances have been extracted by the ordinary operations of Nature. To the left of the archway is a bulky convoluted pillar, rising from the surrounding blocks and boulders to the uppermost part of the portico, and to the right of the archway is a fine piece of stalagmite formation about 10 feet in height. In the centre, immediately behind it, is a large stalactite, and near by an extensive patch which looks like conglomerate of lime and pebbles. On the outer walls are flowering shrubs and creeping plants, including one which bears a strong resemblance to the climbing fig (Ficus stipulata), which clothes with pleasant verdure many an ugly wall in and about Sydney. The rock colouring is especially fine and beautifully shaded all the way from the broad daylight to the beginning of the interior blackness, which is somewhat sharply defined by a fringe of stalactites like the vertical bars of a portcullis.

The immediate entrance to this cave is begrimed with dust. A few yards onward there is an iron gate. The guide opens it and carefully locks in his visitors, who light their candles and proceed by a downward path. The descent is about 80 feet, partly by steps cut zigzag fashion, and then on a sloping floor covered with débris. There is a marked difference in the temperature, which is many degrees higher than that of the outward air, and several degrees warmer than the interior of the Arch and Nettle Caves. Small flies surprise the excursionists by the suddenness of their appearance, and by the narrow limits of their habitat. They live in the zone between daylight and darkness. In the region of perpetual night the only signs of animated nature are clusters of bats. The lighted candles serve to make the surrounding darkness more pronounced.

Where the rays of light pierce through the night to its rocky boundary indistinct, irregular lines can be seen like the ribs of a skeleton, and it is easy to conjure up all sorts of uncanny shapes, from hobgoblins to anthropophagi. The only sounds audible, or apparently audible, are the quickened respiration and the throbbing of the heart. When the voice is raised its effect is strange, and there is no responsive echo. Darkness and silence dwell together. After spending a few seconds—or minutes—in their company, the curator lights his magnesium lamp, and the visitor finds himself in the precincts of "The Cathedral," in the centre of which is a large stalagmite. The roof rises to a height of about 300 feet, 70 feet loftier than Canterbury Cathedral or Notre Dame, and within 100 feet of the altitude of St. Paul's! The walls are composed of limestone, terraced with tier upon tier of stalagmites brought into bold relief by the gloom of innumerable fantastically-shaped recesses. The preacher is Solitude; his theme is "Awful Stillness." Wandering through the nave to the south, the visitor walks over caves not yet opened, but the existence of which can be proved by dropping little pebbles into dark recesses and listening to the percussion on floors more or less remote. In an aisle of the Cathedral leading to the Music Hall, there is another grandly-arched cavern with a steep descent into an abysmal depth. Here on the one side are numerous stalactites, white as virgin snow, and on the other similarly-shaped formations of carbonate of lime tinged with oxide of iron—some of them so deeply as to present the colour of a boiled lobster's crust. This is a favourite clustering place for bats, and numbers of these membranous-winged quadrupeds may be seen snoozing together on the roof.

THE MUSIC HALL.

By means of a wire ladder the excursionist descends still deeper into the bowels of the earth. He then goes farther down by 18 or 20 steps, cut in a clayey substance, to the vestibule of the Music Hall. Some of the stalagmites are stained with clay. They have evidently been used as steadying-posts by visitors who had previously placed their hands on the red earth when working their way down the declivity where the steps are now formed. The other stalagmites away out of reach are white and glistening. The approach to the Music Hall, which was discovered in the summer of 1860, is low. The passage to it is about 35 yards long. The floor is composed entirely of "formation," and at the sides are numerous columns of different colours. The Music Hall itself is about 12 feet in height, and runs out at the end to about two feet. It is called the "Music Hall" because of its very fine acoustic properties. A weak voice raised in song or oratory sounds full and sonorous. This hall encloses a secret which architects of public buildings might covet, and the wonder is how such tonic effects are produced in a chamber which presents so many obstructions to the waves of sound. The floor contains a series of basins, curiously shaped by the water which has been retained in them, until it has escaped by percolation to form stalactites and stalagmites at some lower level. The edges of these shallow reservoirs are sharply defined and gracefully moulded. The formation of the walls is extremely delicate. Some of it is white and some like yellow coral. The roof has been slightly defaced by certain nineteenth century cads. In various places the "mark of the beast," in lampblack, has been produced by holding candles near to the ceiling and moving them about gradually, and the sooty hieroglyphics remain unto this day as an evidence of vanity and folly. The floor, which was once like alabaster, is now soiled by the tramping of feet. But, notwithstanding these defects, the Music Hall is still very beautiful.

THE SHAWL CAVE.

Returning to the main passage, the tourist descends 41 steps, and enters the Shawl Cave, a magnificent chamber, the roof of which slopes at an angle of about 43 degrees. Into one side the "formation" of carbonate of lime has floated like lava in volumes, and presents the appearance of a suddenly congealed cascade. All the adjacent rocks are covered with fine sheets of formation, white and coloured, and hanging in graceful folds. On a far-off wall is more formation of a similar kind, projecting from a perpendicular rock, and variegated with superb tracery and colouring. The "shawls" hang parallel to each other. They gradually increase from six inches to three feet in depth, in a lateral length of from 12 to 15 feet, and at a distance appear as though they had been placed on the wall by an artist; but when the light is put behind them it is seen that they are independent, slightly corrugated, semi-transparent slabs of equal thickness and graduated widths. Of this kind of formation, however, more magnificent specimens are to be found in the Imperial Cave. In another part of this cavern are large detached blocks of formation, which sparkle like diamonds all over the lines of fracture. They are in wild disorder, as though they had been hurled about in some Titanic conflict. The stalactites here are of different character from those found in the other caves, being composite and covered with ornamentation of various kinds. The lower rocks, too, are rippled and chequered like wicker-work, and resemble the formation of the Pink Terraces of Rotomahana, which were destroyed by lava from a volcano in 1886. The roof is about 100 feet high, and the sides of the cave are formed of massive ledges, over which a limey substance has flowed in large masses and assumed elegant shapes fringed with stalactites. Near this place is a hole which goes down to the bottom of another cave. It has not yet been fully explored, but it has been ascertained that its depth is about 120 feet, with a clear pool at the bottom. A stone thrown down it is heard to strike two or three times, and finally splash in the liquid crystal.

CHAPTER XI.

THE EXHIBITION.

At the western end of the Shawl Cave, and on its southern wall, is a remarkable formation denominated "The Butcher's Shop." Experts in the preparation of animal food have discovered in this strong resemblances to sides of beef, joints, and "small goods" covered with a reticulum like the netted membrane sometimes thrown over meat exposed for sale. One would hardly expect to find anything æsthetic about such a display. As a realistic production, however, it will bear favourable comparison with some so-called works of art which show how much humour a jocular sculptor can cut into a piece of cold stone. In its bearing upon gastronomy, exception might be taken to one or two of the joints, which suggest veal that has been "spouted," and an excess of adipose matter; but upon the whole the "shop" may be regarded as a not unpleasing representation of a chamber filled with chilled meat.

Leaving the unromantic stall and ascending seven steps under a roof about 90 feet high, the cave-walker ambulates towards the Exhibition, which is approached by 12 wooden steps, leaving to the right a beautiful formation like a frozen waterfall of from 20 to 25 feet. These steps have pendant from them fungi of the most delicate kind, some resembling eider-down, hanging in flossy masses from underneath the cross pieces. This fungoid growth affords evidence of dampness destructive to the timber, which ought to be replaced by more durable material. It is satisfactory to know that specifications have been prepared and tenders forwarded to the Department for this work. It will be more satisfactory to learn that prompt action has been taken in regard to them, and that they have not been simply docketed and smothered in some obscure pigeon-hole.

The road to The Exhibition is rather rough, there being large masses of angular rocks on either side, and the pathway itself is somewhat rugged. The entrance to the Bride's Cave is to be seen down a rocky declivity of about 30 feet. The gallery leading to this chamber is only about 12 inches by 18 inches. The cave itself is about six feet high, and hung around with drapery of alabaster. The ceiling is of coral formation, and the floor pure white. Farther on to the left is another chamber, the entrance to which is pretty, but difficult of access. It is from 6 inches to 10 feet high. There is beautiful formation in one part from the ceiling to the floor. Some of it is like straws, as clear as glass, and a portion of the floor sparkles as though it were set with diamonds.

The Exhibition is of large proportions, being about 250 feet each way, but its height ranges only from 5 to 20 feet. Its floor is reached by nine steps. From the centre of the Exhibition the entrance to the Bride's Chamber is on the right. To the left is a broken column, which at one time was sound from the floor to the roof, but which has been fractured apparently by the sinking of the rock on which the stalagmitic portion rests. The separation is slight, and there is a slight departure from the right line.

THE BROKEN COLUMN.

To the eastward are several interesting stalactites. One represents a black fellow's "nulla-nulla," another a lady and child, another the palm of a hand blackened by candle smoke. On the south side is a spacious platform like the stage of a theatre—the front, about 40 feet wide, is supported by two columns. The height is about 18 feet, and across the top is a curtain of formation representing drapery gracefully arranged, with a fringe of little sparkling stalactites. On each side of this is a smaller entrance similarly adorned and asexquisitely beautiful. The floor of the stage is about 15 feet deep, and the curved ceiling about 40 feet from the drop curtain to the floor. This is as it appears at a distance. On nearer approach it is perceived that the pillars are uneven, and marked with formations of various kinds. That which seemed like a stage becomes an irregular cavern, with immense rocks lying about in great disorder. When the Exhibition is illuminated by the magnesium light, some beautiful red and white stalactites are disclosed, glittering like dewdrops in the sunlight, and also some exceedingly pretty stalagmites. This chamber was called "The Exhibition" on account of the variety of its specimens. It contains stalactites and stalagmites, white and coloured—variegated shawls—sombre marble and sparkling rocks, clusters of formation, and elephantine masses of carbonate of lime in shapes which prove how much more subtle than professors of art is Nature herself. At the south end a cave slopes down, and there are boulders and débris stained with iron, as well as other indications of great soakage and percolation.

THE JEWEL CASKET.

Eastward, about 40 feet, is the "Jewel Casket." On the way to it are openings to numerous unexplored caves. Affixed to an immense block of limestone are some 30 or 40 shawl-pattern formations of various sizes, which give forth musical sounds when struck with a hard substance, and which, with a little practice, could be played upon like a mammoth harmonicon. En route from the Exhibition to the Jewel Casket, although the passage has not been so dry for twenty years, the rocks are covered with moisture, and the lime can be scraped off like soft soap. From the Exhibition there is a descent eastward of about 100 feet along the gallery, which is somewhat narrow, but the roof of which is covered with pretty stalactites. Near the entrance to the Casket is a remarkable reticulated rock. The descent is by 23 steps east, and then proceeding north about five yards the Jewel Casket is reached.

The Jewel Casket is at the end of a very remarkable cave. Its ceiling is marvellously beautiful. The walls and ridges on each side sparkle like gems of the first water. Some of the rocks are covered with virgin white, and some are delicately coloured. The entrance to the Casket itself is very small, being only about 15 inches by 8. Its upper portion is of glistening rich brown, and slopes in varied graceful folds down to the bed rock. When the magnesium light reveals the splendour of the interior it is seen that the Casket stretches away to a considerable distance; the floor is covered with white and amber brilliants and snowy coruscating flakes of dazzling purity. Here are clusters of cave diamonds, opals, and pearls, with delicate fawn-coloured jewels scattered about promiscuously. Rich and rare are the gems this Casket contains, and exclamations of delight are evoked when their charms burst upon the view like a vision of fairyland. Neither tongue nor pen, nor photographic art nor pencil-sketch, can ever do full justice to this natural treasury of beautiful things.

JUDGE WINDEYER'S COUCH.

Leaving the Jewel Casket, the visitor proceeds in a northward direction along a passage, from the Exhibition to "The Hall to the Bridge." There is an ascent of 13 steps west, and then the way to the Hall is under a low archway, through which it is necessary to proceed on hands and knees. Through this archway is a little cavern, something like the Jewel Casket, with a floor of diamond drift and delicate coral. At the top of the steps the Hall runs north-west. Then the way lies down a gradual slope of rough rocks to the head of 18 steps, with a wire rope on the right hand side. At the top of the steps near to the Jewel Casket and in the Hall to the Bridge is a piece of formation like an upholstered sofa, which has been named "Judge Windeyer's Couch," because it is said that the learned Judge sat on it when he visited the caves. Its surface is of a rich reddish brown, and may have suggested the celebrated woolsack which, in the days of "good Queen Bess," was introduced as the Lord Chancellor's seat in commemoration of the Act to prevent the exportation of wool which was at that time as important an element in England's prosperity as it is at present to the well-being of Australia. In the Hall beautiful formation is seen. A large rock, with shawl-pattern appendages and other ornamentation, is specially attractive. Another represents a miniature Niagara, done in stone. The features are varied by splendid stalactites, from pure white to rich brown. The formation on the wall is like frozen fountains. The bottom consists of huge rocks, angular and rugged, with immense flags of limestone. About 10 yards from the Bridge is "Touch-me-not" corner, with a grotto quite out of reach, but of the interior of which, when the light is flashed into it, a splendid view can be obtained. The stalactites are perfectly shaped and beautifully pure. Some of them are as white as snow, some are opaline, and others are tinged with mineral colours. The floor has many stalagmites and sparkling formations like a jewelled carpet, which falls from the entrance a little distance down the wall in graceful brown folds fringed with russet stalactites. Here the Hall is very spacious, being about 120 feet across, and the roof rises from 10 to 50 feet. It has on it some of the most beautiful stalactites in the caves, many of them being of unsullied white. To the left, high up on the side of the Hall, is a piece of pure lime formation like a lace shawl, the apparent delicate network of which is an object of special interest, if not of envy, to the fair sex.

THE UNDERGROUND BRIDGE.

The Underground Bridge is not a brilliant achievement in engineering, but seems to be well constructed and safe, which is an importantconsideration; for, although it is so many hundred feet below the summit of the mountain, and yet down so low as to be on the same level as the foundations of the Cave House in the adjacent valley, it spans a black yawning gulf, at the extremity of which, 50 feet still farther down, is a clear pool of water 20 feet deep! The Bridge is about 42 feet long. It has wire girders and uprights, with stanchions and handrails, and a wooden deck, which, by-the-bye, needs some repair, for several of the planks are broken. The passage is made increasingly secure by galvanised wire netting stretched along the lower part of the Bridge on both sides. The rocks which form the boundary of the immense chasm spanned by the Bridge are of enormous size, and the scene from this point is remarkable for sublimity rather than for what is commonly called beauty. Near the roof is an immense recess, filled with huge stalactites and mammoth pieces of formation, which have floated over the bottom and formed graceful ornamentation for the cavern below. And so the process is repeated from the top of the immense chamber, near the roof, down to the rugged walls immediately round the Bridge. Even on the rocks which surround the abyss similar wondrous decorations are lavishly bestowed. The clear-headed and sure-footed guide descends from one jutting rock to another and yet another, until he approaches a row of remarkable stalactites which can be just discerned through the gloom. This group is called "The Piano," because of the resonant qualities of its separate parts. Each stalactite gives out a note. The notes vary in quality and pitch, but most of them are imperfect. As stalactites they are very fine, but as melodious instruments they are frauds. They refuse to harmonise, and their music is about as entrancing as that of a discordant "upright grand," mounted on one leg and played with a handle.

CHAPTER XII.

THE LURLINE CAVE.

Seventy or eighty yards from the Underground Bridge is the Lurline Cave. The course is south-west, through a curved gallery with 53 steps in different flights, and two archways—one like loveliness when "adorned the most," and the other formed by an ornate mass of stalactites.

The Lurline Cave is justly regarded as one of the most charming chambers in the group. The coup d'œil is magnificent. It does not need any close examination to find that it has some distinctive features which show that, although there is no aqueous accommodation for the queen of the water nymphs, whose name it bears, the appellation of this portion of the Lucas Cave cannot, etymologically at least, be considered as a lucus a non lucendo. There are the "coral bowers" and cells to which Rudolph was transported; the "halls of liquid crystal, where the water lilies bloom;" there is the cool grot in which the Water Queen dwelt; there is the rock on which she sat "when all was silent save the murmur of the lone wave, and the nightingale that in sadness to the moon telleth her lovelorn tale;" there is Rhineberg's magic cave, with its "wedges of gold from the upper air;" there are the distant recesses to which Lurline sent the gnome while she restored to life her mortal affinity. With such surroundings it is easy to reproduce, link by link, the rosy chain which enthralled the German Count and "The Daughter of the Wave and Air."

Or, to take the more rollicking version by "Thomas Ingoldsby, Esq." Here is "a grand stalactite hall," like that which rose above and about the impecunious "Sir Rupert the Fearless," when he followed to the bottom of the Rhine the dame whose—

"Pretty pink silken hose cover'd ankles and toes;In other respects she was scanty of clothes;For so says tradition, both written and oral,Her one garment was loop'd up with bunches of coral."

Where—-

"Scores of young women diving and swimmingAll slightly accoutred in gauzes and lawns,Came floating about him like so many prawns," and where their queen, Lurline, lost her heart and her plate, and, according to the same reverend author, her cajoler, whose disastrous fate inspired the moral—

"Don't fancy odd fishes! Don't prig silver dishes!And to sum up the whole in the shortest phrase I know,Beware of the Rhine, and take care of the Rhino!"

The floor is covered with hemispherical mounds or domes for the naiads to recline on. The outer wall is composed of formations ranged in festoons of stalactites—not smooth and transparent, but opaque white, and marked with all the wonderful elaboration which characterises zoophytic work in the coral reefs of the Southern seas. This cave contains several sub-caves, each of which has special charms, and the turning of some of the arches is marvellously graceful. One of the recesses is filled with stalactites which look like groups of seaweed. The coral is russet and cream colour and saffron, and there are honeycombed rocks varying in shade from vandyck brown to chrome yellow. Some of the stalactites in the interior sub-caves are transparent. Whichever way the eye is turned it encounters submarine grottoes of fantastic shape, decorated with imitations of algæ. If it were only at the bottom of the Rhine instead of thousands of feet above sea-level, it would seem natural as well as beautiful, but here its existence is simply a wonder, and the sensation produced is fairly described by the last word in the marriage service of the Church of England. Still, "when Mother Fancy rocks the wayward brain," it is easy to associate with it denizens of the deep, and people it with naiads, or with Undines, who were supposed to marry human beings, and, in certain conditions, become endowed with human souls. The cave is about 15 feet high, and from 15 to 20 feet broad. Some of the coralline ledges at the sides are remarkably handsome, and many of the stalactites are from six to eight inches in diameter. The cavern is elegant in its proportions, highly favoured in regard to stalactite growth, graceful in contour, and rich in colouring.

THE FOSSIL BONE CAVE.

About 15 yards north-west from the Lurline Cave is the Fossil Bone Cave. To reach this cavern it is necessary to ascend 12 steps. It is scarcely less beautiful than the Lurline Cave. The lime formation represents pensile boughs of weeping-willow, garlands of flowers, and stalactites covered with all kinds of floral decorations. Here also are some fine "shawl" formations hanging from the rocks. One of them is called "The Gong," because it produces a sonorous note similar to that of the Chinese instrument which is superseding the dinner-bell, and challenging its title to be regarded as "the tocsin of the soul." On a sloping side of the floor are some forms distinctive in shape and colour, and resembling a lot of small potatoes shot down indiscriminately. The wonder is how in such a place they could have been so formed and isolated. Here is an oblique cavern, at the bottom of which a bone of some large animal lies embedded in the limestone formation like a type in a matrix. At one time it was doubted whether this, which appeared to be bone, was really an osseous substance, but subsequent examinations have proved that it is bone. A fracture of the rock has shown that the outer part of the bone is compact, and the inner part cellular. It is beautifully white, and, as the formation about is brownish, the phosphate can be readily distinguished from the carbonate of lime. On the roof above the Fossil Bone Cave is a rare stalactite about 20 feet in length, and by the side of the tomb of the unknown animal—which may have been anything from a diprotodon to a dingo—is a splendid monumental stalagmite. The cave is about 50 feet high, and 50 feet in length and breadth. The roof is of a light cream colour, and has brown stalactites of perfect shape. The side rocks are magnificently draped. Numerous splendid columns like white marble, and sheets of stalactitic growth, excite wonder and admiration.

THE SNOWBALL CAVE.

About 40 yards through a hall, running north-east of the Fossil Bone Cave, is the Snowball Cave, which is about 9 feet high, 25 or 30 feet long, and from 6 to 10 or 12 feet wide. It runs north-north-east. Its distinctive feature is that its roof and a portion of its walls are covered with little white masses like snowballs. Some of the patches of carbonate of lime stick to the walls in isolated discs, and others are massed as though snowballs had been thrown at a mark, and a number of them had stuck close together. Some of the stalactites in this chamber have been formed by the upward pressure of water, and assume many tortuous shapes. An interesting feature of this portion of the caves is the existence of a number of stalactites which show how readily vibration is communicated from one to another. The visitor puts his finger to the end of a stalactite, and when an adjacent one is struck so as to make it sound, it is perceptible that the vibration of the sounding stalactite is communicated to its silent neighbour.

There is one more chamber to visit in the Lucas Cave. To reach it the visitor ascends four steps, and travels north-west about 14 yards to the head of a wire ladder, which he descends to a place directly underneath the Snowball Cave, and then he goes down the steps into the Wallaby Bone Cave, over the entrance to which is a very pretty cluster of stalagmites, from 6 inches to 18 inches long, and varying from the thickness of a straw to three-quarters of an inch in diameter. The floor is covered with wallaby bones, and in the immediate vicinity are quantities of osseous breccia.

CHAPTER XIII.

THE BONE CAVES.

The Bone Caves are intensely interesting, and a considerable amount of attention has been paid to them by scientists. In 1867, Professor Owen, when writing to the Colonial Secretary, said that the natural remains obtained from the limestone caves of Wellington Valley in 1832, "revealed the important and suggestive fact that the marsupial type of structure prevailed in the ancient and extinct as well as in the existing quadrupeds of Australia." Seventeen years ago there was an expedition to the Wellington Valley Bone Caves. Parliament voted £200 for the purpose, and an investigation was made by Mr. Gerard Krefft, who at that time was curator of the Australian Museum, and Dr. Thompson. They obtained many valuable and rare specimens, some of which were said to be quite new to science, consisting of the remains of mammals, birds, and reptiles. The largest bones and teeth discovered were of a size equal to those of a full-grown elephant. They were remains of diprotodons and nototheriums, gigantic marsupials now extinct.

The Wellington Valley Caves were discovered by Sir Thomas Mitchell more than 50 years ago. From them no fewer than 2,100 specimens of fossil remains were presented to the British Museum. When the result of the exploration was forwarded to Professor Owen, he said that the conclusion was very much what might have been naturally looked for, and that the only disappointment he felt was the absence of human remains and works. Ten years ago an attempt was made to obtain the co-operation of the neighbouring colonies in the work of thoroughly exploring the caves of the western and southern districts and Australian rivers. The proposition originated with the Agent-General for New South Wales, Professor Owen, and Sir George Macleay, but the adjacent colonies did not see their way to participate, whereupon our Cabinet decided to do the work without extraneous aid, and £600 was voted by Parliament for the service of 1882. At an earlier stage Professor Liversidge had written to the Colonial Secretary, transmitting the following extract from a letter he had received from Professor Boyd Dawkins, M.A., F.R.S., of Owens College,

Manchester:—"Would the Government of New South Wales undertake the systematic exploration of the wonderful caves which are in the colony, and which certainly ought to be explored? Not only is there a certainty of adding to the great marsupials which have been obtained, but there is a great chance of finding proof that man was living at the same time as the extinct animals, as he has already been found in Europe and Asia. I should expect to find a very low form of the aborigine. Such an inquiry would be of a very great interest to us here in England, who are digging at the caves all over Europe, and the duplication which would be obtained would enable the trustees of the Australian Museum to increase their collections largely by exchanges."

The minutes of the meetings of the trustees of the Australian Museum show that in 1881 a committee, consisting of Dr. Cox, Mr. Wilkinson, and Professor Liversidge, was appointed for the management of the exploration of caves and rivers, and it was decided that the following caves should, if possible, be examined in the order as written:—Wellington Caves, Cowra, or Belubula Caves, Abercrombie, Wollombi, Fish River (now Jenolan), Wombeyan, Wallerawang, Cargo, Yarrangobilly, Murrumbidgee, Kempsey. The Goodradigbee caves were also included, and from them was taken a great quantity of bones of small animals, with a number of jaw, thigh, hip, and shin bones of some animals of the kangaroo family. The smaller bones were those of mice, bats, birds, and marsupials. In the Wellington breccia cave a shaft was sunk, and on the 20th September, 1881, Mr. E. P. Ramsay, curator of the Museum, reported, among other things, the following:—"A great number of interesting bones have already been obtained from this shaft, but the mass of 35 feet of bone breccia which we passed through shows that we have here a large field for exploration. From this shaft we have obtained bones of the following animals, besides a great number of small bones yet undetermined—Diprotodon, macropus, palorchestes, sthenurus, procoptodon, protemnodon, halmaturus, thylacinus, bettongia, sarcophilus, phascolomys, dasyurus, phalangista, pteropus (?), bats, rodents (mus), a few lizards' bones, and a few vertebræ of lizards and snakes."

Other caves also were explored, but it was found that the bones obtained from them were of recent origin. It is a question whether it would not be desirable to make still further investigations. The osseous breccia—where it exists—appears to be similar in all the caves. There are rifts and pits and chambers where animals have retired to die, and where from time to time their bones have been formed into cement with the liquefied rock, which in process of time has again hardened and become a solid compound of bone and stone.

In the southern room on the first floor of the Sydney Museum is a large collection of bones from the Wellington and other caves. These remains have been collected during the last four or five years under the direction of Mr. Ramsay, the curator. They are chiefly the bones of marsupials. There are not among them any fossil remains which indicate the presence of man in Australia at any very remote period. Some of the principal bones are those of extinct marsupials, and are important from a scientific point of view. They include bones of the following animals (species extant) found in the Wellington caves:—The thylacinus (Tasmanian tiger), sarcophilus (Tasmanian devil), mastacomys (a rodent), hapalotis albipes, and mus lineatus (New South Wales). Other important fossil remains in the Museum are those of the thylacoleo (two species), diprotodon, procoptodon, protemnodon, palorchestes, macropus titan, nototherium, phascolomys. There are not in the Sydney Museum any bones from the Jenolan Caves—which, however, contain many interesting remains of the animal world,—because the search for them would involve the destruction of attractive features. For these reasons attention was given to the Wellington Caves, whose beauties were not likely to receive further disfiguration than they have already suffered.

From the Wallaby Bone Cave the visitor returns to the Fossil Bone Cave, and ascends a wire ladder which is about to be replaced by an iron staircase. As he mounts this wire-rope ladder, which is 76 feet long and not "stayed," he feels the necessity for some better means of communication. From the top to the Cathedral is about 25 yards south-east. A large portion of the cave north-west from this point has not been explored. There are five or six different branches, one of which runs out to daylight at a small aperture (14 inches by 18 inches) over the rise of the water below the Grand Archway and the Devil's Coach House. The distance from here through the Cathedral to the entrance gate is about 70 yards, up two flights of steps. There is a gradual ascent to the steps, and the final flight of 41 brings the excursionist to the gate and to the sunshine. He will be glad to rest awhile before entering the Imperial Cave, which is the grandest of them all.

CHAPTER XIV.

THE IMPERIAL CAVE.

The Imperial Cave is graced with myriads of lovely objects. Darkness brooded over them for ages, as drip by drip and atom by atom they were formed into things that charm and shine in chambers whose walls are "clad in the beauty of a thousand stars." There are underground gullies terrible enough to be the home of Apollyon, with legions of goblins; and strangely radiant elfin palaces where Titania might be supposed to reign, and Robin Goodfellow carry on his frolicsome pranks. In the year 1879, when the cave-keeper (Mr. Wilson) discovered this magnificent series of caverns, he was lowered down a distance of 90 feet through Egyptian darkness. As this mode of access was neither cheerful nor easy, nor free from danger, he determined, if possible, to find a less inconvenient and perilous approach to the cave. After two years of patient investigation he accomplished his heart's desire. The orifice which has been converted into the present entrance was at first, for a distance of 19 feet, only 14 inches by 15 inches, but the curator worked his way through it, caterpillar fashion, with a light in one hand and a hammer in the other, knocking off the rough formation, and widening the aperture from time to time until he made communication free from difficulty. Throughout this splendid cave there are many places where similar efforts, accomplished with equal success, have added largely to the safety and convenience of visitors, who reap the fruits of the heroic work performed by the brave explorer, whose best years have been spent in rendering accessible to the public the marvellous beauties of the Jenolan Caves.

From the accommodation house the way to the Imperial Cave is through the Grand Arch, on the northern side of which, at the eastern end, are two wooden staircases. The first springs from the floor of the arch amidst immense blocks of stone irregularly disposed. It has 21 steps, and a handrail on each side. This terminates at the summit of a pile of limestone rocks, the uppermost of which forms a platform guarded by iron stanchions and a galvanised wire rope. From this platform there is another flight of 21 steps to the portico of the cave—a plain archway, the floor of which is 50 feet higher than the floor of the cave-house. The entrance is guarded by a light iron gate.

THE WOOL SHED AND THE GRAVEL PITS.

About 35 yards from the entrance to the Imperial Cave, northward, and thence about 30 yards east, is "The Wool Shed." The approach to it is narrow and low. In some places it has been formed by blasting, and in others by excavation through a red, sandy substance underneath the limestone. It widens as the Wool Shed is approached. In the floor is a hole going down to the former entrance to the cave, now closed by a stone wall. The Wool Shed is about 20 feet wide, 15 feet high, and 70 feet long. The formation over a large part of the walls and roof resembles the fleeces of sheep, hanging about and spreading over the shelving rocks in all directions. There is one pelt which suggests the "Golden Fleece" torn by Jason from the tree trunk in the poison wood guarded by the huge serpent spangled with bronze and gold, and which was soothed to slumber by the magic song of Orpheus. The surroundings are as strange as those of the lonely cave where dwelt Cheiron the Centaur, who taught the leader of the Argonauts "to wrestle and to box, and to hunt, and to play upon the harp." But perhaps, after all, it may be only an indifferent limestone representation of a fellmongering establishment. The woolly skins and scraps are mirrored on the retina. The impressions produced by the sense of vision depend not upon the optic nerve, but upon the imagination. Simply as a spectacle, however, the Wool Shed is curious and entertaining. The blocks of stone near to the base are for the most part plain, and the floor is broken and rugged.

Descending 12 steps, and passing through a tunnel five feet six inches by two feet, the visitor stands at the junction of the right and the left hand branches of the cave. Here formerly the passage was only 14 inches by 15 inches. The larger opening was made by blasting, and the material blown from the solid rock has been packed away in recesses at the side of the hall, which, at the junction of the two branches, widens out considerably, but does not present any specially interesting features. The right hand branch runs north-west, and the left hand branch runs south-west. Taking the south-west branch first, after travelling about 10 yards the visitor comes to "The Gravel Pits," which he reaches by ascending a mound with 13 steps. There are two pits of gravel. One of them is about 12 feet deep and the other about 15 feet. In the rocks overhead are bones distinctly visible, owing to the earthy matter having fallen away from them. Some of these bones are large. There are shelving rocks about six feet from the floor. The sides of one of the Gravel Pits are oblique, but the other pit, which is railed off, is round and perpendicular. It could hardly have been more symmetrical had it been made by a professional well-sinker. This spot, although perhaps uninteresting to a mere sight-seer, cannot fail to attract the attention of geologists. Ascending two flights of stairs with 14 steps each, the excursionist attains a height of about 40 feet above the Gravel Pits in a north-westerly direction. Between the two flights of steps the ground is sloping, and the walls hold a considerable portion of drift, the pebbles of which are large and tinged with oxide of iron. This passage leads to the Margherita Cave, and from it a tunnel branches off towards the "Architect's Studio." This is a very pretty vestibule, about 30 yards in length, and bearing south-east. At first it rises several feet by steps, and later on there is a descent of five steps through masses of stalactites, and past a beautiful pillar.

THE ARCHITECT'S STUDIO.

The height of the "Studio" is about 18 feet. This atelier is a marvel of beauty. There are in it two temples of the most lovely kind. Large masses of splendid stalactites hang from the roof. On the walls are columns profusely decorated with coral and tracery and bosses, and carvings which could be imitated only by the most cunning workmanship. Near the centre is a large stalactitic mass, most graceful in shape, with numerous appendages; and underneath appear several stalagmites. Some of them have been partially destroyed, but one, which touches the enormous mass of stalactites above, remains intact. Near to this is a splendid column, richly embellished. The walls are profusely adorned with elaborate configurations, which are supposed to represent architectural "studies," from which the cave derives its name. Most of the formation is white or light grey; but in some of the recesses there is rich colouring. Each chamber has its own distinctive attractions, and contains many objects which challenge special admiration. Massive grandeur is set off with the most delicate and fragile beauty. Stalagmites are not numerous here, but one about eight feet in height, and two inches in diameter at the base, tapers off gradually towards the roof until it becomes as attenuated as the thin end of a fishing-rod. The stalactitic formation hangs in ponderous grotesquely-shaped concretions, some of which extend from the roof nearly to the floor, and many of the stalactites which decorate the stalactitic formation are perfect in shape and purity. The choicest portions of the Architect's Studio are fenced off with galvanised wire rope on iron standards.

THE BONE CAVE.

Ascending a flight of 10 steps out of the Architect's Studio the course is south-west about 30 yards to the Bone Cave. The way is difficult, a portion of the journey having to be performed on hands and knees. The cave, which runs north and south, is about 10 feet high, 150 feet long, and from 5 to 30 feet wide. In the middle of it is a passage only partially explored. The Bone Cave is guarded by iron rods and wire netting. Bunches of stalactites hang from the roof, and the floor is strewn with bones, covered with a thick coating of lime formation. There are also bones embedded in the floor. Some of the formations on the floor are very peculiar, consisting of small curiously-shaped pieces fitted together at remarkable angles, and yet capable of being taken to pieces like triplicate kernels pressed together in one nutshell. A large proportion of the stalactites are quite transparent and decorated with small sharp points, and some formations among the coral are as lovely as fine marine mosses, which they resemble. In the midst are numerous unexplored recesses, which, when the light penetrates, are seen to hold hundreds of fine stalactites, crystal and opaque. The objects of beauty in the Bone Cave retain their colour, because they cannot be handled by that class of visitors who fancy that they can see only with their fingers. On the walls are specimens of delicate fretwork, and on the floor as well as on the top of rocky ledges, stalagmites lavishly ornamented. Although not as grand as the Architect's Studio, this is a very fine cave, and additional interest attaches to it in consequence of the fossil bones it contains. The adjacent chambers cannot be explored without destroying some of the well-known beauties of the cavern.

CHAPTER XV.

THE MARGHERITA CAVE.

From the Bone Cave to the Margherita Cave is about 130 yards, travelling north-east to the top of the first 10 steps, then east into the Architect's Studio, and then north about 30 yards. The Margherita Cave varies from 10 to 20 feet in height, and is from 10 to 15 feet wide. It is remarkable chiefly for the magnitude and beauty of its stalactitic formation, the best portions of which are fenced off with iron rods and wire netting. The formations are nearly all of the same general character. Although there are many changes in detail, the typical pattern is observed everywhere in the midst of infinite variety, just as in a fugue choice snatches of melody sound forth in the clear treble, skip away in the mellow tenor, roll forth in the deep bass, and then dart about Will-o'-the-wisp-like all through the composition, without ever getting out of harmony. It is a grand chamber full of stately concords and charming effects of light and shade.

Hard by is another chamber with masses of beautiful stalactites, and, on a pinnacle, a figure appears about the height of the Venus de Medici, robed in drapery of white, slightly suggestive of the binary theory of feminine attire, and with a peculiar curvature denominated the "Grecian bend." The bend is unmistakable. There is just a suspicion of the "divided skirt," and the attitude is easy and graceful, the Grecian bend notwithstanding. The upper part of the body from the waist has no "boddice aptly laced," but becomes gradually mixed indiscriminately with other kinds of beauty, which, although they may "harmony of shape express," do not in the sense indicated by Prior become "fine by degrees and beautifully less." Admirers of classic beauty may be inclined to regard the incompleteness of the figure as "fine by defect and delicately weak." There are some stalagmites on the sloping bank of formation, which runs down to the wire netting and is finished off at each extremity by two massive stalactitic pillars.

The Margherita Cave received its name in honour of the wife of Lieut.-Colonel Cracknell, Superintendent of Telegraphs. Col. Cracknell visited the caves in 1880, and on the 22nd July illuminated this and some other portions with the electric light. The Margherita was the first of the underground chambers in which flashed its brilliant rays.

In the absence of facilities for generating electricity by means of the now well-known dynamo machine, Colonel Cracknell had recourse to primary batteries, and adopted the form known as the Maynooth or Callan cell, the elements of which were cast iron and zinc in solutions of nitric and sulphuric acid.

It was not an easy task to unload and carry up the iron cell battery and the apparatus into the cave, as each set of six cells weighed 96 lbs. The whole, together with the acids and the electric light apparatus, exceeded 15 cwts. The battery, however, was soon made ready, and to the admiration of all present Cave Margherita was illuminated by the electric light. A photographic apparatus was then placed in position, the plates were exposed, and in 15 minutes the first negatives were produced, and said to be all that could be desired.

It is satisfactory to learn that arrangements are almost complete for the permanent lighting of the caves by electricity. Lieutenant-Colonel Cracknell proposes to illuminate them in sections, containing each, say, 25 incandescent lamps, and when one section has been thoroughly explored the lamps therein will be cut off and those in the next section brought into operation, and so on until the whole of the interior has been examined. It is intended that Swan's incandescent lamp of 20-candle power shall be used.

The electricity is to be generated by a small Edison dynamo, with which accumulators of the Elwell-Parker type will be kept charged, so that at all times there will be a supply available for lighting the lamps. It has not yet been determined whether to use steam or water power, but it is thought likely that sufficient of the latter may be secured in the vicinity of the caves to work a turbine, and thus produce the necessary energy.

CHAPTER XVI.

THE HELENA CAVE.

Leaving the Margherita Cave by a descent of five steps, and travelling north-west about ten yards through a festooned hall, the Helena Cave opens to view. It was named in 1880. Helena is the prenomen of Mrs. Hart, whose husband accompanied Lieutenant-Colonel Cracknell on his visit to the caves, and took photographs of some of the chambers, when for the first time they were illuminated by electricity. Mr. Hart was connected with the photographic branch of the Government Printing Office. The pictures then produced, although large and fairly good, are not equal to some more recent photographs taken when the chambers were illuminated by the magnesium light.

The Helena Cave is about 60 yards long, 15 to 20 feet high, and varies in width from 20 to 50 feet. For stalactitic splendour it will bear comparison with the most magnificent of the caves. There are columns like the trunks of stately trees, covered with rough formation resembling coarse bark. Coralline masses droop laden with myriads of cells. In the recesses are stalactites perfect in shape—crystal, and alabaster set off by others coloured like ferruginous sandstone. Lovely grottoes and decorated rock ledges abound. In one or two instances joined stalactites and stalagmites form pillars with bunches of formation all about them like stony efflorescence. Several steps lead into a recess, the floor of which contains basins made by the action of water.

The formation throughout is remarkable for its lavish ornamentation and purity. Among the grand cornices is one weighing about half a ton, formed in such a manner as to resemble great bunches of grapes, like those brought from Eschol by the Hebrew spies to illustrate their report on "the promised land." In other parts are small clusters like vine produce growing en espalier. It seems as though in these subterranean sunless bowers nature had by some subtle process striven to reproduce in stone the fruits and flowers of the sunned surface, clothing them in pure white and sombre grey, and endowing them with charms as sweet and mutely eloquent as the fragrance of the Cestrum nocturnum, or the cold beauty of a night cactus bloom which caresses the moonbeams or wantons in the stellar light.

This place, full of enchanted grottoes and elfin palaces, gives, perhaps, the best illustration of the plan, so uniform and yet so diverse, on which these limestone mountains have been honeycombed into galleries, "high overarch'd with echoing walks between," and caverns large and small, from cathedral spaciousness to the minimised dimensions of the tiniest chamber in the finest coralline structure. Their infinite gradation may be fairly described by certain well-known lines, and substituting the word "caves" for the name of the most lively insects of the genus pulex—

Big caves have little cavesAnd lesser caves about 'em;These caves have other caves,And so ad infinitum.

The most remarkable feature hereabouts is a piece of formation called "The Madonna." It is supposed to represent a woman carrying an infant, which rests on her right arm. The left arm hangs loosely by her side, and the right knee is bent as in the act of walking. The head bears less resemblance to that of one of the favourite creations of the Old Masters than it does to the anterior part of a Russian bear. A pyramidal mound about four feet high forms a pedestal for the figure, which is about two feet six inches from crown to sole. A sculptor with mallet and chisel might in an hour or two convert it into a representation of loveliness, but at present it is only a veiled beauty. Visitors have to imagine all those witcheries and feminine perfections portrayed by great artists who have made "The Madonna and Child" a life study.

The best view of this cave is that looking south-east with "My Lady" in the centre. The stalactites show to great advantage, and as the manifold charms brought into bold relief by the magnesium light disappear, and the sable goddess "from her ebon throne, stretches forth her leaden sceptre," the sensation produced is one of pleasant bewilderment. The deep gloom which follows celestial brightness enshrouds the glorious scene. The pageant fades away as did the celebrated palace which Potemkin reared for his Imperial Mistress. It was made of blocks of ice. The portico was supported by Ionic pillars, and the dome sparkled in the sun, which had just strength enough to gild, but not to melt it. "It glittered afar like a palace of crystal and diamonds, but there came one warm breeze from the south, and the stately building dissolved away until none were able to gather up the fragments." So it is with these underground wonders. They are brought into bold relief, and gilded by the brilliant light of the magnesium lamp. It is extinguished, and the gorgeous palaces and solemn temples suddenly become like "stuff which dreams are made of."

Another beautiful feature in the Helena Cave is a formation under a mass of stalactites which hang from the roof and drop water on to a jutting rock below. On a corner of this shoulder is a huge epaulette, and underneath are some elegantly-shaped brackets. Still farther down is an enormous richly decorated mass, flanked by shell pattern formation. The base rock rests on a mound of limestone gracefully curved, and the intervening spaces are filled with myriads of ornate specimens. Some distance above the floor is a bold rock with a sharply cut under-surface like the sounding-board of a pulpit hung with stalactites. Here are also terraces like miniatures of the celebrated White Terraces of New Zealand, with basins, the sides of which are graced with a formation which at one time was pure white, but the lower portions of which are now discoloured. The upper part, however, still retains its pristine purity and loveliness. The terraces approaching The Grotto are stained by the tramping of feet. About halfway up is a handsome stalagmite of fine proportions. This chamber is grandly impressive, and remarkable for its charming variety of formation, as well as for its graceful contours.

CHAPTER XVII.

THE GROTTO CAVE.

At the point of exit from the Helena Cave there is a descent of four steps. Then it is necessary to ascend 14 steps north-north-east on the way to the right-hand branch of the Imperial Cave. From the top of the steps the distance to the junction is about 80 yards. On the left side of the passage, at the foot of the lower steps in the left-hand branch, is a drive down into the gallery of the right-hand branch, the fall being about 70 feet. It was by being lowered down this hole that the cave-keeper found that portion of the right-hand branch which extends from the shaft to the junction of the two branches. This perilous part is railed off with two wires supported on iron standards let into the rock. At a point 22 yards north, on the passage to the Grotto Cave, at an angle, is a drop of 100 feet into the right-hand branch of the Imperial Cave.

Sitting on a thin shell of limestone, on the right-hand side, the visitor may pitch a stone into a hole 10 inches by 14 inches, and hear it strike the bottom of the black depth. He may thrust his candle down to arm's length underneath the mineral crust, and (if he be in a very cheerful vein) fancy he is peering into the Infernal Regions, over which he rests on a thin and fragile screen. From this point the Grotto Cave is south-south-west about 50 yards. Precautions have been taken against accident at this spot. Iron standards are let into the rock, and there are double wires stayed back to the walls of the cave. It is intended to make it still more secure on the lower side by a netting of three inch wire, on one and a quarter inch iron standards, from four to six feet high.

Where the rock has been cut to make the passage wide enough, the steps are wet, and there is a little basin always full of bright water of a bluish tint. A false step here might precipitate a sight-seer into an almost perpendicular hole, some idea of the depth of which may be formed by casting a stone down, and listening to its striking against the sides, until after the lapse of several seconds the sound of its contact with the floor rises like a feeble voice, still further subdued by distance. Descending five steps, a good sight is obtained of the Grotto Cave. It is 25 feet high in places, and about 10 feet wide, with passages in all directions. It runs south-south-east and north-north-west, and is full of interesting vaults and crypts, over which Nature seems to have cast a mystic spell. For alluring charms, fantastic combinations, and disposition of matter, no comparison can be found between it and the most artistic grottoes built by human hands. One grotto is roofed with delicately white and richly-traced formation, studded with stalactites of rare splendour. Here is a delicate white shaft piercing the dome; there a stalagmite within half-an-inch of the stalactite above. A broken pillar suggests either an accident or a barbaric act; near to it is a perfect column, which, in the dim light, seems like a figure emerging from the cave. Close inspection reveals imitations of coral and seaweed, curved stalactites, and filagree work of the most intricate design. Little flakes of lime, like snow, at the back of the grotto, sparkle like twinkling stars.

Another grotto, in the centre of the cave, is made entirely by large stalactites, set off with small ones. Some are covered with filaments about the thickness of ordinary sewing-cotton; others seem as though they were covered with beautiful mosses. Many of the pendants are richly wrought and extremely graceful. The upper stalactites are covered with thicker filaments like twine and pack-thread. A third grotto is remarkable chiefly for a splendid cornice or buttress projecting from a pillar. It is as grand, though not as ornate, as similarformation in the Margherita Cave. It was named on the 10th March 1881, and its designation is appropriate.

Near the exit is a marvellous grotto, at the entrance to which is a massive stalagmitic pillar, five feet in diameter, meeting the end of a stalactite about 15 feet long. The back of the stalagmite constitutes a separate grotto of stalactites and shell-pattern formation. Near to it is a remarkable rock, covered with cauliflower-shaped masses of limestone. It is known as the Cauliflower Rock—the choufleur of the gnomes who guard the unfathomable caves of Jenolan. In yet another grotto, at the rear of the main pillar, is a beautiful canopy, with thin stalactites, straight like walking-canes, and others thin as straws, crystal and opaque. There are also many contorted stalactites and other eccentricities in stone. A little iron ladder makes it easy to descend into this cave of so many beautiful grottoes on gracefully undulating foundations. Near the point of departure is a dangerous spot, for the proper guarding of which arrangements are being made. Adjacent is a considerable quantity of red clay covered with smooth white formation, and fractures of rock round about sparkle with crystals.

CHAPTER XVIII.

THE LUCINDA CAVE.

After travelling a few yards south from the grottoes the visitor arrives at the Lucinda Cave. The hall is from 5 to 15 feet wide and from 6 to 25 feet high. The approach to the cave is through limestone rocks. The path has a gentle slope, and in some places the walls are besprinkled with a white substance like wool. Near the centre of the passage is a hole in the roof partly lined with formation which sparkles like a starlit sky. A little farther on is a descent of four steps through a passage, the walls of which glitter with great brilliancy.

About 25 yards from the Lucinda Cave, south-south-west, is a magnificent spectacle. The roof is densely crowded with stalactites of every type of beauty. On each side are crevices of dazzling splendour, and on the floors of which brilliants have been showered like hail. The largest remain in the centre, and the corners and other remote places are heaped with diamond drift. In one place in the lower cave is a bank made of formation washed from the hall above, thickening to the base at an angle of about 40°, and studded with cave gems. Between the jewelled floors and the superincumbent rocks are stalagmites of pure white calcareous alabaster. At the end of this passage are three steps, which it is necessary to ascend in order to reach the Lucinda Chamber, which was discovered on the 7th February, 1881, and is named after the wife of the curator of the caves.

The Lucinda Cave is from 10 to 25 feet high, from 50 to 70 feet wide, and about 90 feet long. Its entrance junctions with the steps to Katie's Bower. To reach the Lucinda Cave from the junction, the visitor passes over a level floor, like cement, about 12 feet in length. This cave is remarkable for its scenic grandeur. The spectator stands in mute admiration, and gazes upon the magnificent sight like one who is spellbound. The beauty is Brobdignagian in its proportions. The figures are all colossal. There are immense stalactites and stalagmites of every hue. An enormous mass of formation droops from the ceiling to the summit of a stalagmitic mound upon which it rests. It is like a series of suddenly congealed waterfalls, and the groundwork below is gracefully rippled on the outer surface, and fringed with stalactites. The mound previously mentioned rests upon another of larger size, of equally graceful contour, and besprent with brilliants which sparkle like immense diamonds. To the right is a cascade of formation which has trickled and solidified from rock to rock and from ledge to ledge in graceful curves from the roof to the floor.

Those who have seen water arrested by congelation on an extensive weir, and rendered opaque by hoar, can form a tolerably correct idea of the kind of beauty here represented in stone. To the left of the frozen waterfall is a bower of sparkling substances, and at its extremity is a recess, from the farthest visible point of which can be seen magnificent clusters of stalactites, of rich and varied colouring. This bower is more chastely beautiful than any ever possessed by Oriental potentate. The walls on the left side are richly draped with sheets of formation of uniform thickness, hanging from the roof like shawls or scarves. This mineral drapery is opaque, striped and flecked with russet and reddish brown, and edged with white as pure as virgin snow. It is guarded by a fence of iron rods and galvanised wire; consequently it is impossible to make a close and minute examination of its interior, but the general effect is fascinating. In one of the recesses is a terraced rock covered with reddish formation, like a cascade, which certainly is not less beautiful than were the Pink Terraces of Rotomahana. A little beyond the cascade the same kind of formation ornaments a massive pillar, which constitutes one of the principal features of the cave.

In the foreground is a hall which leads to an unexplored region below, and the entrance to which is guarded by a fence to prevent accidents. The floor is curiously formed by a series of basins, the rims of which are shaped into every variety of curve and indentation, running in and out like frilling, not with regular curved lines like escallops, but representing in miniature the waterlines of a quiet harbour with large bays and pretty inlets and creeks and reaches, without a single straight line. The edges of these basins are about two inches in height, covered on the outside by sparkling limestone, like delicate coral, thickening towards the floor. Inside the formation is still more beautiful, with coralline matter of the same general character swelling out to the most graceful concavity. There is perfection in every segment, and in every tiny cell lurks tremulous light.

CHAPTER XIX.

KATIE'S BOWER.

Taking a course between the parallel fences of wire-netting, and travelling about 25 yards, the Jewel Casket is seen on the western side of the cave. It extends about 20 feet due south, and is about two feet six inches high by about six feet wide. It does not contain any new type of beauty, but rather represents a collection of the most enchanting cave splendours. Even the stalactites and stalagmites are overlaid with ornamentation. It is a focus of dazzling brilliancy.

Returning to the junction, there is a descent by steps south-west into "Katie's Bower," which is the last chamber in the left-hand branch of the Imperial Cave. Forty-three of these steps are like cement, 10 are cut in the solid rock, and 19 are of wood. On the left-hand side, at the foot of the staircase, is a hole 70 feet deep. To the north of the entrance, at the top of the wooden steps, is a remarkable formation suggestive of a Chinese pagoda, waxy-amber-and-flesh-coloured. To the left is a beautifully-formed dome, with convexities of pure white, sparkling all over as though it were studded with diamonds cut with large facets. At the bottom of the dome are numerous stalactites, resting on a curved rock coated with the same material. It is like a richly bejewelled throne with a grand canopy suspended from the roof by a large stalactite. It contains many hundredweight of formation, and is about three yards from end to end of the curve. It is ornamented with filagree work and stalactites of the most curious and beautiful kinds, and on the upper surface are some handsome stalagmites. At the rear of this splendid canopy, set with precious stones, is a recess with a sparkling floor.

Having descended some steps to the south-east, and ascended 30 others through a broken part of the subterranean region, the visitor will find much to admire in Katie's Bower. It is about 250 feet long, 5 to 30 feet high, and from 15 to 25 feet wide. Its direction is from north-west to south-east. The north-west end is very rough, with a rocky floor. The beauties of the Bower are located to the south and south-east. There are on the one side alabaster pillars, on the other is formation. Immediately over the arch is a deposit of red clay, which has imparted a rich colour to the huge stalactites which hang from the roof. The light of the candles is flashed back by glittering crystals. The floor, which forms the entrance to the Bower, is carpeted with glistening alabaster. Descending 14 steps into the Bower there is a fountain full of lime-water, and a plate suitably inscribed conveys the information that Katie's Bower was discovered on the 7th February, 1881, by Jeremiah Wilson (guide), C. Webb, H. Fulton, C. West, J. Bright, E. Webb, E. T. Webb, J. Thompson, W. H. Webb, E. Bowman, W. Thompson, J. McPhillamy, R. Thompson, J. Webb, and S. Webb. The before-mentioned gentlemen were the first to enter the Bower after its discovery. They had rendered valuable assistance to the guide, and were well rewarded for all their trouble by the consciousness that they had participated in opening to the public a new and charming scene in this western wonderland. The stalactites and formation at the mouth of the cave are pure alabaster.

It should be here stated that this Bower was named in honour of a daughter of the Hon. E. Webb, M.L.C., of Bathurst, who at various times has interested himself in regard to the caves, and sister of the Messrs. Webb whose names appear on the tablet, and who supplied ladders and ropes to the curator, and otherwise assisted him in his explorations. It is a grand cavern, with massive pillars and large stalactites, and elaborate alabaster structures, more remarkable for richness of detail than the most wonderfully-constructed Oriental temple. The dome commences near the roof with a conical mass of brilliant formation, from which depend many fine stalactites, which rest on a solid mass, and seem to hold it suspended. This second mass of formation is ornamented with stalactites like frozen jets of water. And so the process is continued again and again, until the points of the lowest stalactites rest on a white mass level with gracefully-curved and coloured rocks, which descend with elegant contours to the bottom of the Bower. The same kind of wonderful fabrication is repeated at the sides of the principal figure. Some of the flooring is as rich and pure as that of the Jewel Casket. It is a marvel of intricate grandeur, and has the advantage of having been well preserved. From the alabaster stalagmite in front, to the most delicate lime drapery on the walls, there is no prominent "mark of the beast." The cads of the period have not been permitted to perform their favourite ceremony of the laying on of hands.

It would be useless to speculate as to the time the caves in this branch "took in building." It defies all calculation. Apparently the process of formation is finished. There is no dripping from the stalactites. There may be, however, in wet weather; but it seems as though the creative action had given way to the hardening process. It is suggestive of that portion of "King Solomon's Mines" in which H. R. Haggard has a clever and somewhat caustic conceit respecting stalactitic growth. On his way through the enormous cave leading to Solomon's Treasure-house, he was enchanted with the gigantic pillars, which looked like ice, and which sprang in lofty and yet delicate beauty sheer to the distant roof. "Others again," he says, "were in process of formation. On the rock floor there was in these caves what looked exactly like a broken column in an old Grecian temple, whilst high above, depending from the roof, the point of a huge icicle could be dimly seen. And even as we gazed we could hear the process going on, for presently with a tiny splash a drop of water would fall from the far-off icicle on to the column below. On some columns the drops only fell once in two or three minutes, and in these cases it would form an interesting calculation to discover how long at that rate of dripping it would take to form a pillar, say 80 feet high by 10 in diameter. That the process was, in at least one instance, incalculably slow, the following instance will suffice to show. Cut on one of these pillars, we discovered a rude likeness of a mummy, by the head of which sat what appeared to be one of the Egyptian gods, doubtless the handiwork of some old-world labourer in the mine. This work of art was executed at about the natural height at which an idle fellow, be he Phœnician workman or British cad, is in the habit of trying to immortalise himself at the expense of Nature's masterpieces, namely, about five feet from the ground; yet at the time that we saw it, which must have been nearly 3,000 years after the date of the execution of the drawing, the column was only eight feet high, and was still in process of formation, which gives a rate of growth of a foot to a 1,000 years, or an inch and a fraction to a century." This is a very good satire upon the presumption of some modern disciples of the illustrious Cocker. A botanist may, by its concentric zones, tell the years of an exogenous plant; a bucolic sage may judge the age of horned cattle by counting their horny rings; but to tell the æons of a stalagmite is more difficult than the accurate compilation of a feminine census. Arithmetical calculations on such a subject would probably be received with as much confidence as phrenological evidence of the character and habits indicated by bumps on the head of the Great Sphinx at Ghizeh.

CHAPTER XX.

THE RIGHT-HAND BRANCH OF THE IMPERIAL CAVE.

Having thus completed his inspection of the left-hand branch of the Imperial Cave, the visitor returns to the junction, passing through all the chambers previously noticed in it excepting the Architect's Studio and the Bone Cave, and proceeds to explore the still more wondrous and beautiful works in the right-hand branch.

THE SUBTERRANEAN RIVER.

The first object of interest in the right-hand branch of the Imperial Cave is the Subterranean River, which runs at the bottom of a fearful chasm about 50 yards from the point where the two branches bifurcate. After having wandered through marble halls and crystal palaces, and bowers where "rural fays and fairies dwell," the course seems rather gloomy. But attention is attracted by some curiously-shaped nodules, like those found in a part of the Arch Cave, and by basins with thin laminated sides slightly corrugated. These specimens reveal the secret of the construction of the pretty reticulated mounds, which give such a charming effect to several of the most regularly formed features of the caves. To complete the process, the sharp parallel lines which form a succession of little equidistant ridges require only to be smoothed off by a gently flowing film of water, and to receive a coat of colouring derived from clay or oxide of iron. This part of the branch, therefore, should not be passed through hurriedly, for it is instructive.

The road is fairly good, although the arching is low. Those parts, the narrowness of which formerly made progress difficult, have been enlarged, but a pretty natural bridge has been carefully preserved. The halt is at the end of a wire ladder bent over a cliff, which forms one side of an immense gulf, where perpetual darkness broods. Here the visitor has a more ungraceful task to perform than that of the cursed serpent, for he must recline face downwards and "progress backwards" until he assumes the form of an obtuse angle, with one line over the precipice. Then he has to use his legs pretty much as an octopus uses its tentacles, to gain a footing on the ladder, which descends about 50 feet on the chasm side of the angle.

Having found the first rung he feels happy, but not sufficiently hilarious to slide like a lamplighter. He grips the side wires carefully, takes heed to his steps, and goes down slowly. When he has descended a little way, the dim candle-lights above appear to be far distant, and when he is 20 or 30 steps down they look like glowworms. The journey, however, is not perilous to persons who possess a fair share of agility and nerve. It is frequently performed by ladies, of whom the guide is specially careful, preceding them and keeping just below them on the ladder. As this is, perhaps, the most interesting of all the cave sights, it is desirable that some easy mode of descent should be provided, such as a skip like those employed in mines, or a lift made by machinery to work as easily and effectively as those which ascended and descended at will in the subterranean world inhabited by "the coming race." It was, perhaps, some such place as this which suggested to Bulwer Lytton the chasm down which his nameless hero descended to the bottom of an abyss illuminated with a diffused atmospheric light, soft and silvery as from a northern star; where he found lakes and rivulets which seemed to have been curbed into artificial banks, some of pure water, and some which shone like pools of naphtha; where the birds piped in chorus, and where he made the acquaintance of the An people and the Gy-ei, who moved through the air without effort, who had for servants automata always obedient, and totally ignorant of the eight-hours system, and whose religion had these two peculiarities: "Firstly, that they all believed in the creed they professed; and, secondly, that they all practised the precepts indicated by the creed."

Underground rivers appear to be natural to limestone caves. The reason of their existence is to be found in the fact that the mountains in which they are formed are, in geological parlance, "dykes." They must not be confounded with old river beds, such as are encountered by miners—where the surface of the earth has been raised by deposits of alluvium, or where the geological condition of things has been changed by volcanic action. These cave rivers have all been formed by water finding the lowest attainable level in its passage to the sea, and by the solid limestone rocks which have barred its direct course, and have been undermined by its subtle but persistent action. The fluid, dammed back by the mountains, has simply obeyed the laws of gravitation and accumulated force, as evidenced in the trickling silvery thread which follows the course of ant-tracks; in the laughing rill which makes its bed among the pebbles; in the babbling brook which leaps to the swelling river; and in the mighty torrent whose strength and velocity proclaim the majesty of hydraulic power. In all parts of the world where limestone dykes and caves exist, it is reasonable to expect to find subterranean rivers. The eye of the seer can follow the water drips—

"Down through caverns and gulfs profound,To the dreary fountain-headOf lakes and rivers underground."

He can see them again when the rain is done—

"On the bridge of colours seven,Climbing up once more to heaven,Past the setting sun."

But the underground rivers found in caves perform vagaries outside the sweet imaginings of the poet and the prevision of the seer. Farfrom the beaten track they turbulently force their way through recesses and tunnels and pockets of the earth, before they are again warmed with sunshine, and glow in the harmonious colours which form the Bow of Promise.

The Rev. Richard Taylor, in his "Te ika a maui," refers to interesting caves near Mokau (New Zealand), in some of which bones of the moa have been discovered. About a mile from Pukemapau he came to a limestone range, and entered a large cave called Tanaureure. At the bottom of a chasm he found a fine crystal spring, about a foot or so deep, but appears not to have been particularly inquisitive as to whence the water came or whither it went.

A little distance up one of the tributaries of the Rewa River, in Fiji, is a crystal streamlet which flows on towards a lofty ridge, near to which it sinks into the earth. At the mouth of a dark cavern can be heard the roaring. It is a grand expansive excavation, but

"Dark as was chaos ere the infant sunWas roll'd together, or had tried his beamsAthwart the gloom profound."

The water rushes through narrow chasms as through a race, collects in a large pool, and flows through a distant outlet, marked by a speck of light, like a tiny star.

At the Weathercote Cave, in Yorkshire, a stream swallowed up by a rocky mouth is thus described by Walter White in his book entitled "A Month in Yorkshire":—

"The rocks are thickly covered in places with ferns and mosses, and are broken up by crevices into a diversity of forms, rugged as chaos. A few feet down, and you see a beautiful crystalline spring in a cleft on the right, and the water turning the moss to stone as it trickles down. A few feet lower, and you pass under a natural bridge formed by huge fallen blocks. The stair gets rougher, twisting among the big, damp lumps of limestone, when suddenly your guide points to the fall at the farther extremity of the chasm. The rocks are black, the place is gloomy, imparting thereby a surprising effect to the white rushing column of water. A beck running down the hill finds its way into a crevice in the cliffs, from which it leaps in one great fall of more than 80 feet, roaring loudly. Look up: the chasm is so narrow that the trees and bushes overhang and meet overhead; and what with the subdued light and mixture of crags and verdure, and the impressive aspect of the place altogether, you will be lost in admiration.

"To descend lower seems scarcely possible, but you do get down, scrambling over the big stones to the very bottom, into the swirling shower of spray. Here a deep recess, or chamber, at one side, about eight feet in height, affords good standing ground, whence you may see that the water is swallowed up at once, and disappears in the heap of pebbles on which it falls."

In the Wombeyan Caves, near Taralga, in this colony (New South Wales), there is a similar phenomenon. The mountain in which the caves are formed dams, at right angles, a valley of considerable length. On the one side a creek flows into a hole underneath the "Wombeyan Church"—a name given to the principal entrance. This shallow hole is filled with large boulders and less bulky water-worn stones, through which the water instantly filters and disappears. The suction is perceptible if a hand or foot be placed in the basin. On the other side of the mountain the water, running at a considerable distance below the surface, can be seen through an orifice. Farther on, about three-quarters of a mile from the mountain side, the stream bubbles up like a fountain, and reminds the visitor of antique pictures representing the rush of water from rock-smitten Horeb.

The most gigantic of underground river wonders are to be found in the Mammoth Cave of Kentucky, where the Echo River is navigable for three-quarters of a mile, where the Roaring River raises its liquid voice, where Lake Lethe soothes to forgetfulness, and where there is a veritable Styx with a nineteenth century Charon. But although everything about that cave is colossal, it cannot surpass Jenolan in its deep emotional effects, or in the admiration it evokes. Indeed, in these respects an American visitor, familiar with the Mammoth Cave, has given the palm to Jenolan.

Having descended the 50-feet ladder, the foot of which is clamped to a bare rock, the excursionist watches the guide hopping gleefullydown, with candles in one hand and paraphernalia in the other. He then peers into the darkness to find the river, and is led along a gradual slope of about ten feet, when he comes to its margin without recognising it, and would possibly soon be up to his knees in it if he were not warned by his cicerone. It is apparently motionless, as smooth as a mirror, and so clear that at first it is difficult to believe there is any water there at all. You see the sand and pebbles and rocks at the bottom, but do not perceive the fluid which covers them.

The river is about 10 feet wide, and from 12 to 18 inches deep. The length visible is about 60 yards. The overhanging rocks range from 5 or 6 feet to 70 feet in height. From one end of the river comes a sweet soothing sound, made by water-friction. This proves the existence of a considerable current, but the fluid is so limpid and smooth that the eye cannot detect its motion. It is demonstrated, however, in another way. The curator cuts little sections of sperm candle, and, after lighting the wick, floats them on the river, whereupon they immediately begin to glide down the stream, the course of which is nearly south-east. The effect is extremely pretty. Not only are the lights themselves sharply mirrored below, but there is a perfect reflection of the rocks above. Near the tunnel by which the water emerges N.N.W. is a mass of overhanging formation, duplicated in a natural mirror. The bed of the watercourse is dark, being covered chiefly with mud and grit and a few water-worn pebbles. The rocky walls are of limestone—white and black. Up the channel N.W., about 40 yards, is a good crossing-place—not in old Charon's boat, but by means of an ordinary deal plank. On the other side of the river is a ledge of rocks with pebble drift concreted with a substance somewhat resembling the cement in which diamonds are found, and the pebbles, although larger, are of the same shape and blackness as those commonly associated with the luminous gems found in their natural state at Kimberley, in South Africa, or in the Tenterfield district of New South Wales.

On turning the light of the magnesium lamp up the river, its rocky ledges are seen to be ornamented with stalactites and formation perfectly mirrored in the water, which is about 600 feet from the surface, and about 50 below the level of the Cave House in the centre of the valley. Never had river more romantic barriers. Human imagination could not conceive a freak of nature more wildly grand or mysteriously beautiful. There are large ornamented pillars near delicately-tinted formation, drooping from overhanging rocks, like pensile boughs of weeping willow. Some of the twigs skim the surface of the stream, and others are bathed in it. Beyond is a water-hole about 40 feet long, and from 16 to 20 feet wide. Because of its wonderful clearness, it is difficult to judge of its depth; but it has been tested to the extent of five feet, and probably at the extreme point where the water flows from the tunnel it may be six or seven feet deep. The effect of the brilliant light is superb. The ornamentation on the roof of the tunnel is reflected and transposed in the mirror below, each reflected stalactite having the appearance of a twin stalagmite rising from the river bed, which may be traversed for about 150 yards.

Nearly six months ago the caretaker placed in this river twenty young carp from Bathurst. Some of them were enticed from their cavernous resting-places by the bright rays of the lamp, and appeared to be tolerably vivacious and in fair condition. They seem to have all they require except the solar rays; but what is life without sunshine! They ought to be scientifically observed, for there is a theory that in three generations of darkness they will become blind. This has been the fate of the fishes in the Mammoth Cave of Kentucky, and it is stated that their blindness is the result of a law of Nature, which does not continue to supply organs or powers which have ceased to be necessary. Dr. Forwood, in his history of the Mammoth Cave, says:—"The fishes are of a peculiar species, and are of a class known as viviparous, which give birth to their young alive, and do not deposit eggs after the manner of most other fishes. They have rudiments of eyes, but no optic nerve, and are, therefore, incapable of being affected by any degree of light.... It has been proved that these eyeless fishes prey upon each other. In shape they somewhat resemble the common catfish, and rarely exceed eight inches in length." Professor Silliman published the following in his "Journal" for May 1851:—"Of the fish there are two species, one of which has been described by Dr. Wyman in the American Journal of Science, and which is entirely eyeless. The second species of the fish is not colourless like the first, and it has external eyes, which, however, are found to be quite blind. The crawfish, or small crustacea, inhabiting the rivers with fish are also eyeless and uncoloured; but the larger-eyed and coloured crawfish, which are abundant within the caves, are also common at some seasons in the subterranean rivers, and so also, it is said, the fish of the Green River are to be found in times of flood in the rivers of the Caves." Dr. Forwood gives also the following quotations, on the authority of Professor Agassiz, an eminent naturalist in the department of ichthyology:—

"The blind fish of the Mammoth Cave was for the first time described in 1842 in the 'Zoology of New York,' by Dr. Dekay, part 3rd, page 187, under the name of 'Amblyopsis spelæus,' and referred, with doubt, to the family of 'Siluridæ,' on account of a remote resemblance to my genus Cetopsis. Dr. J. Wyman has published a more minute description of it, with very interesting anatomical details, in vol. xlv. of the 'American Journal of Science and Arts,' 1843, page 94.

"In 1844, Dr. Tellkampf published a more extended description, with figures, in 'Müller's Archiv' for 1844, and mentioned several other animals found also in the Cave, among which the most interesting is the crustacean which he calls 'Astacus pellucidus,' already mentioned, but not described, by Mr. Thompson, President of the Natural History Society of Belfast. Both Thompson and Tellkampf speak of eyes in these species, but they are mistaken. I have examined several specimens and satisfied myself that the peduncle of the eye only exists; but there are no visible facets at its extremity, as in other crawfish.

"Mr. Thompson mentions, further, crickets, allied to 'Phalangopsis longipes,' of which Tellkampf says that it occurs throughout the Cave. Of spiders, Dr. Tellkampf found two eyeless, small white species, which he calls 'Phalangodes armata' and 'Anthrobia monmouthia'; flies, of the genus 'Anthomyia'; a minute shrimp, called by him 'Triura cavernicola'; and two blind beetles; 'Anophthalmus tellkampfii' of Erichson, and 'Adelops hirtus;' of most of which Dr. Tellkampf has published a full description, and figures in a subsequent paper, inserted in Erichson's 'Archiv,' 1844,.

"The infusoria observed in the Cave resemble 'Monas Kolpoda,' 'Monas socialis,' and 'Bodo intestinalis,'—a new Chilomonas, which he calls 'Ch. emarginata,' and a species allied to 'Kolpoda cucullus.'

"As already mentioned, Dekay has referred the blind fish, with doubt, to the family of Siluridæ. Dr. Tellkampf, however, establishes for it a distinct family. Dr. Storer, in his 'Synopsis of the Fishes of North America,' published in 1846, in the 'Memoirs of the American Academy of Arts and Sciences,' is also of opinion that it should constitute a distinct family, to which he gives the new name of 'Hypsæidæ,' page 435. From the circumstance of its being viviparous, from the character of its scales, and from the form and structure of its head, I am inclined to consider this fish as an aberrant type of my family of Cyprinodonts."

The effect of long-continued darkness upon visual organs has had some remarkable illustrations. At one time an idea prevailed in America that caves possessed certain curative properties, and afflicted people remained in them; but the absence of light proved disastrous to many. It is recorded that those patients "who remained in the cave three or four months presented a frightful appearance. The face was entirely bloodless, eyes sunken, and pupils dilated to such a degree that the iris ceased to be visible, so that, no matter what the original colour of the eye might have been, it soon appeared black."

This subterranean river offers a fine opportunity for scientific observation well worthy to be embraced by some Australian naturalist. In the vicinity of the river is to be noticed one of the few signs of decay to be found in the caves—a portion of shell pattern formation shows evidence of mouldering, and appears like a mere skeleton. When the visitor has ascended the ladder and safely negotiated the angle at the top, he feels that he has witnessed the most interesting place to be found in the western wonderland; and when he fishes for a compliment to his agility, and is reminded of the graceful forms that occasionally ascend and descend in much better style, he immediately recalls a Patriarch's dream, and thinks the ladder ought to be named after Jacob.

CHAPTER XXI.

THE FOSSIL BONE CAVE, THE SPARKLING ROCK, AND THE CRYSTAL ROCK.

About 20 yards north from the ladder to the underground river is the entrance to the Fossil Bone Cave. Here is a stratum of coffee-coloured slatey substance in layers like those of the Wianamatta shale. It is so soft that a gentle touch is sufficient to pulverise it. It is slightly honeycombed, and its outer surface is covered with imitations of delicate lichens. In this cave there is not anything in the shape of a stalagmite, except an empty brandy bottle on a little shelving rock, and that would be generally regarded as a bad substitute. For æsthetic as well as for other reasons, it would be better to keep such "stalagmites" out of the caves. On the top of a large rock is a mass of "washdirt," 14 inches deep, with "headings" of about the same dimensions. Some experienced diggers say they never saw more promising stuff. During the yellow fever from which so many suffered a few years ago, companies were floated on the strength of "claims" equally delusive. Proverbially, "auriferous ground" is deceitful, and this "washdirt," which looks rich enough to make a prospector's eyes sparkle with delight, has proved to be as worthless as a lying prospectus. A portion of it was washed, but did not show the colour of gold. It remains, however, an object of interest, and may serve to teach a useful lesson.

There are in this cave solid limestone rocks above and below. The roof is about 500 feet beneath the surface of the mountain. In the bed of the cavern are many fossil bones. Some appear to be remains of native dogs. In various rocks are clusters of bat bones. A very noticeable osseous object is the vertebra of a bird with one side-bone. There is no trace of the other. There are also many large bones, the cylindrical cavities of which are filled up with formation. Some of these bones are deposited about eight feet from the bottom of the cave. On one ledge is a heap of bones, large and small. Some of them are very fine specimens. The height of this cave is about 15 feet, and its breadth from 8 to 15 feet.

Travelling from the passage leading to the Fossil Bone Cave N.N.W. about 40 yards, and passing through a cutting N.W. about 30 yards, the visitor comes to "The Sparkling Rock." A cutting, five feet by two feet, forms the entrance to a spacious hall, where is seen the Sparkling Rock, large and shelving. The principal portion of it is slightly coloured, but the lower part is beautifully white. It is hung with large stalactites and fleecy pendants. Some of the formation resembles sheepskins, with the woolly side outwards, thrown negligently over the ledges. From this point the course runs west about 30 yards through a hall from 20 to 25 feet high, and from 15 to 20 feet wide, and thence north about 40 yards to the Crystal Rock.

About 14 yards west from the Sparkling Rock, and by a road wide enough for a coach and four, there is a large upward shaft of about 100 feet to the Grotto Cave, which, as previously stated, is between the Helena and the Lucinda caves in the left-hand branch. Here the tourist gets a good idea of the way in which the two branches of the Imperial Cave are situated with regard to each other. The right-hand branch is the lower series. The left-hand branch is higher up in the interior of the mountain, and to the south-east, with the exception of the Grotto Cave, which is immediately overhead, and about 100 feet from the Sparkling Rock.

On the left-hand side of the passage, and about 25 yards from the Crystal Rock, is a very pretty grotto of formation, with an overhanging ornamental mass like a canopy. Up above, about 40 feet, is the opening to an unexplored cave, the mouth of which is composed of solid shining rock, with white stalactites. There are also, round about, coloured stalactites varying in length from an inch to a couple of feet. The remainder of the passage is lofty and rugged.

Not far from the entrance to the Crystal Rock is the bottom of the shaft down which the curator was lowered from the Coral Cave (a sub-cavern of the Elder Cave) into the Imperial, and on the wall this memorable event is duly recorded. Here we read:—"These caves were discovered by Jeremiah Wilson." Then follows a list of the names of persons who lowered the fearless curator down the black hole: "Alfred Whalan, Thomas A. Gread, Jeremiah F. Cashin, Joseph Read, Nicholas Delaney, Ralph T. Wilson, Thomas Pearson, Heinrich Neilzet, and William Read." They were named "Wilson's Imperial Caves" on February 16, 1879. From this spot the

Sparkling Rock is about 15 yards N.N.E. It is about 25 feet wide and about 18 feet high. Stalactitic formation descends from an angle in the roof, and rests on four or five finely coloured terraces which glitter all over as though they were covered with spangles. To the left of these terraces is a large basin with coral sides and a rim composed of three or four layers of shell-shaped pattern overlapping like fish scales, the rows being a little way apart from each other, and the intervening spaces filled with formation. The bottom of the basin is covered with very delicate ornamentation, deposited by water which has soaked through to a lower level. In the background is another rock, covered with similar formation, fringed with stalactites, and stalactites also descend to it from the roof.

CHAPTER XXII.

THE SHAWL CAVE.

About 30 yards from the Sparkling rock is the Shawl Cave. It is approached through a passage from six to eight feet high and two to four feet wide, containing numerous small but pretty grottoes. The Shawl Cave is very interesting. To the left of the entrance is a grotesque pillar with little domes of snowy whiteness and masses of stalactite. The cave is about 25 feet long, 15 feet high, and from 12 to 15 feet wide. It contains three magnificent "shawls." One is 14 feet long, 18 inches deep, and one-sixth of an inch thick, and in the blending of colours represents tortoise-shell. The other two are not quite as large as the first-mentioned. They are straw-coloured, varied with rich brown. They hang at right angles from the side of a convex sloping roof, and the colouring runs from end to end in parallel lines, but the bands of colour vary in depth. For instance, the first piece of the shawl—say one inch and a half from the roof—may be pure white formation, of lime, or carbonate of lime coloured with oxide of iron which gradually becomes paler and paler. The next two inches may be light yellow, spotted with brown. The next strip may be fox-colour, and so on, until the design is completed. For the most part, the cave "shawls" are of uniform thickness, like sheets of opaque glass slightly corrugated transversely. The opposite wall is nearly perpendicular. At each end of the cave is a grotto. One is low down and gloomy-looking; the other lofty, going up into the roof and full of formation, some of which is like frost work. The stalactites are immense. From the further wall are sloping terraces, gradually enlarging towards the base underneath the hanging shawls. There are also some remarkable clumps of formation. One is like a giant's foot; another resembles the skull of a wolf, or of some other animal related to the canine tribe.

About seven yards north from the Shawl Cave is a cavern 20 feet broad, 30 yards long, and from 12 to 14 feet high, the principal object in which is "The Lady's Finger." Under a shelving rock fringed with stalactites of all the prevailing colours, and almost every variety of shape, the "finger" forms the extremity of a stalagmite about 12 inches high, and similar in figure to a feminine forearm in a sleeve, with coral trimmings. The forearm is white, and the chubby hand is of a waxy-looking flesh colour. The thumb and the index finger point upward. According to the Talmud, "man is born with his hands clenched, and dies with his hands wide open;" in reference to which one of the Rabbinical sages remarks—"Entering life he desires to grasp everything; leaving the world, all that he possessed has slipped away." This hand with the lady's finger, however, is not grasping, and it points upwards. The modern science of chiromancy, according to A. R. Craig, M.A., in his interesting book "Your Luck is in Your Hand," divides hands into seven classes: "1. The hand elementary, or hand with a large palm; 2. The hand necessary, or spatulated; 3. The hand artistic, or conical; 4. The useful or square hand; 5. The philosophical, or knotted hand; 6. The psychological, or pointed hand; 7. The mixed hand." It would be difficult to class the hand with "the lady's finger" in any of the foregoing divisions, and it would puzzle one skilled in palmistry, and who regards the human hand as a mirror of the mind, to use it even in the way phrenologists use the casts of bull-necked, animal-headed felons. The index finger is long, the pollex (thumb) is short; the medius (middle) is wanting, and so are the annularis (ring finger) and the auricularis (little finger), "so named by the Romans because of its utility in cleansing the ear." The visitor, therefore, must not expect to find here a hand like a model of perfection on a Greek statue; but he will see a remarkable alabaster extremity, sufficiently well formed to be called "the lady's finger." The rocky bank, which is coloured with several shades of brown, and veined with formation, is also flecked with white, like snow. At one end of the cave the view closes with long-sparkling stalactites—those nearest being brown and flesh-coloured. Behind them is pure white formation which sets off to great advantage the beautifully-tinted stalactites sparsely scattered about the cave. The other end of the cavern gradually tones off to sombre rocks of grey and brown.

At the end of the Lady's Finger Cave is a charming grotto, and, above, the rocks are like fine coral in various shades of red and grey. Inside the grotto are stalagmites thick at the base and with elegant stalactites resting on them. Some are pure white, and others are covered with fine tracery. In front is a perfect stalactite which descends to within an inch of a perfect stalagmite just underneath it, and aptly illustrates the process of their growth. On the floor are pretty hillocks of somewhat dismal-looking matter which, on close inspection, is seen to be made of coralline figures and sparkling crystal atoms. In the foreground is a fine stalagmite, fitted all over with minute coral. This group, protected by wire netting, is specially interesting because it is unblemished. All round the approaches are little bunches of stalactites like epaulettes.

After travelling west about nine yards, ascending five steps, and then proceeding 14 yards north, the tourist arrives at a cave containing a very conspicuous column called "Lot's Wife."

CHAPTER XXIII.

LOT'S WIFE.

The alabaster pillar called "Lot's Wife" stands in solitary grandeur within a gloomy cave. Its sombre surroundings are in harmony with the tragic old-world history recalled by the central figure. The nimble thought skips over ages and ages, and in the "mind's eye" appear the rich plains of Siddim and the flowing Jordan, and the fugitives and the lava, and the terrible climax. As the Biblical record of the catastrophe is supposed to teach the folly of disobedience on the part of wives, and the perils of hankering after doubtful pleasures, the pillar which recalls it may be contemplated with advantage by newly-married couples, now that the caves are becoming a favourite resort of honeymooners. Perhaps in time to come there may be religious services and solemnisation of matrimony in these fantastic subterranean caverns. It is related by Dr. Forwood, that a romantic marriage took place in the Gothic chapel of the Mammoth Cave of Kentucky, "which family interference prevented occurring on the earth." He says: "The fair lady, whose lover was opposed by her parents, in a rash moment promised them that she would never marry her betrothed 'on the face of the earth.' Afterwards, repenting of her promise, but being unable to retract and unwilling to violate it, she fulfilled her vow to her parents as well as to her lover by marrying him 'under the earth.'"

How far the pillar in the caves is like that mentioned in the Book of Genesis it is impossible to say, because the latter has been neither minutely described nor photographed. Josephus, the great historian of the Wars and Antiquities of the Jews, and who was not born until about 2,000 years after Lot's departure from Sodom, says he saw it. His words are: "When Lot went away with his two maiden daughters—for those who were betrothed to them were above the thoughts of going, and deemed that God's words were trifling—God then cast a thunderbolt upon the city, and set it on fire with its inhabitants.... But Lot's wife continually turning back to view the city as she went from it, and being too nicely inquisitive what would become of it, although God had forbidden her to do so, was changed into a pillar of salt." And, he adds, "for I have seen it, and it remains to this day." It is to be regretted that he did not describe the pillar itself.

A century later Irenæus bore testimony to the existence of the pillar, and spoke of its lasting so long "with all its members entire." This would lead to the inference that the original pillar retained the shape of a female figure. If it did, then in this respect there is no similarity between the Pillar of Warning on the Dead Sea plain and the pillar in the Jenolan Caves. The latter is a pretty round column, about five feet four inches high, rounded off irregularly at the top, and built up in sections, which show separate growths, like divisions in the stem of a cabbage-tree palm, or the joints of a bamboo. It is probable, therefore, that there is not the slightest resemblance between the two pillars. Bishop Patrick thinks that some of the storm which overwhelmed the Cities of the Plain overtook Lot's wife, "and falling upon her as she stood staring about, and minded not her way or guide, suddenly wrapped her body in a sheet of nitro-sulphurous matter, which, congealing into a crust as hard as stone, made her appear, they say, as a pillar of salt, her body being candied in it."

It is about 3,800 years since the disobedient "help-meet" of the Oriental squatter was fixed like a fly in amber, as a solemn warning to recalcitrant spouses for all time. Had the first drip then fallen on to the mound in the Jenolan Caves where now stands "Lot's Wife"? Query. The Jenolan pillar is evidently of slow growth. Each joint, which looks something like fine tallow, may, as the curator facetiously puts it, represent a century of "dripping." In this respect it is unlike the historic pillar whose name it bears. Dr. Kitto, in his very interesting "Daily Biblical Illustrations," says in reference to the latter: "From the nature of the case, and from the peculiarly bituminous and saline character of the locality through which this phenomenon was produced, we must not expect to discover many parallel instances which might be quoted in illustration. Accordingly we find that the illustrative parallels which have been diligently sought out by the old commentators have rarely any real bearing on the subject, being for the most part accounts of people frozen to death and long preserved in that condition uncorrupted in the Boreal regions, or else of persons suffocated and then petrified by the mineral vapours of the caves in which they were hid, or otherwise of persons 'turned to stone,' and found generations after standing in the postures wherein they found their death. The only instance we have met with which seems appropriate, and which rests on the authority of a contemporary of fair credit, is related by Aventinus, who states that in his time about 50 country people, with their cows and calves, were, in Carinthia, destroyed by strong and suffocating saline exhalations which arose out of the earth immediately upon an earthquake in 1348. They were by this reduced to saline statues or pillars, like Lot's wife, and the historian tells us that they had been seen by himself and the Chancellor of Austria."

It was, perhaps, some such incident as this which gave to Mr. Haggard the idea as to how the Kukuana people from time immemorial preserved their royal dead. He first of all described Twala, the last of the Kukuana kings, as in a limestone cave, with his head perched upon his knees and his vertebræ projecting a full inch above the shrunken flesh of the neck. "Then," he says, "the whole surface of the body was covered by a thin glassy film caused by the dripping of lime-water. The body was being transformed into a stalactite." The antecedent kings were ranged around a table in this wonderful cave, and the author continues:—"A look at the white forms seated on the stone bench that ran around that ghastly board confirmed this view. They were human forms indeed, or rather had been human forms; now they were stalactites [stalagmites?]. This was the way in which the Kukuana people had from time immemorial preserved their royal dead. They petrified them. What the exact system was, if there was any, beyond placing them for a long period of years under the drip, I never discovered; but there they sat, iced over and preserved for ever by the silicious fluid. Anything more awe-inspiring than the spectacle of this long line of departed royalties, wrapped in a shroud of ice-like spar, through which the features could be dimly made out (there were 27 of them, the last being Ignosi's father), and seated round that inhospitable board, with Death himself for a host, it is impossible to imagine. That the practice of thus preserving their kings must have been an ancient one is evident from the number, which, allowing for an average reign of 15 years, would, supposing that every king who reigned was placed here—an improbable thing, as some are sure to have perished in battle far from home—fix the date of its commencement at four and a quarter centuries back. But the colossal Death who sits at the head of the board is far older than that, and, unless I am much mistaken, owes his origin to the same artist who designed the three colossi. He was hewn out of a single stalactite [stalagmite?], and, looked at as a work of art, was most admirably conceived and executed." There is nothing suggestive of anything so hideous as this in the Jenolan Caves. "Lot's wife," as she appears there, is as straight down as a "Shaker," without the slightest suspicion of artificial "improvement." Nor does the pillar correspond with the result of more recent discovery made by an American expedition to the Dead Sea, and in reference to which Dr. Kitto says:—"The course of their survey could hardly fail to bring under notice every marked object upon either shore, and one they did find, an obviously natural formation, which—or others in former times like it—might readily be taken by persons unaccustomed to weigh circumstances with the precision we are now accustomed to exact, for the pillar of Lot's wife. Among the salt mountains of Usdum (an apparent transposition of Sodom), on the west side of the kind of bay which forms the southern extremity of the Dead Sea, the party beheld, to their great astonishment, while beating along the shore, a lofty round pillar, standing, apparently detached from the general mass, the head of a deep, narrow, and abrupt chasm. They landed, and proceeded towards this object over a beach of soft slimy mud, encrusted with salt, and at a short distance from the water, covered with saline fragments and flakes of bitumen. The pillar was found to be of solid salt, capped with carbonate of lime, cylindrical in front and pyramidal behind." The italics are the Doctor's. It is not novel to say that history repeats itself; but it is questionable whether among the fashionable inhabitants of the Cities of the Plain in the days of Lot the modern crinolette was a feminine artifice of that Worthless time. According to the Koran, Lot's wife, Waila, was in confederacy with the men of Sodom, and used to give them notice when any strangers came to lodge with him "by a sign of smoke by day and of fire by night." In this regard the pillar at Jenolan may be regarded as a warning, and not as suggestive of anything, except, perhaps, the lesson conveyed by the Apocrypha, in the Book of Wisdom x. 7, where there is a reference to Lot's wife, "Of whose wickedness even to this day the west land that smoketh is a testimony, and plants bearing fruits that never come to ripeness; and the standing pillar of salt is a monument of an unbelieving soul." Is it not a pity that so beautiful a column in the most wonderful caves ever made by Nature should have been associated with so much that is off-colour? True, it is itself a little crooked and irregular, but these characteristics are accounted for by its peculiar formation. It has not been produced in the ordinary way by drippings from one stalactite, but, contrary to rule, owes its origin and development to two small stalactites in the roof. Consequently, its growth has been continually warped. It is, however, a beautiful feature of the Imperial Cave, and may teach many useful lessons to persons of observation and nous.

CHAPTER XXIV.

THE CRYSTAL CITIES—THE SHOW-ROOM AND THE GRAND STALACTITES.

From "Lot's Wife" to "The Crystal Cities" is about 20 yards north, through a hall from 9 to 15 feet high. On the right-hand side is a concrete wall, which rises about 12 inches from the floor, to protect the "Cities" from dust raised by the tramping of feet. At the end of this concrete wall is a descent of two steps, which brings visitors in full view of the exquisitely beautiful cave, in which there is a group of dazzling Lilliputian cities, whose buildings are of crystallized lime. The streets appear to be thronged with minute figures

"— — no bigger than an agate stoneOn the forefinger of an alderman."

The sight recalls the Man Mountain and the wonderful Land of Lilliput, upon which Lemuel Gulliver was cast, where cavalry exercised on the palm of his hand, and infantry marched 24 abreast between his legs, which were stretched out like those of a colossus. Imagination can supply the Palace of Belfaborac in the metropolis of Lilliput, surrounded by myriads of tiny statuettes, representing the kingdom in which raged no less than six rebellions, excited by an imperial decree that eggs should be broken only at the smaller end, whereas it had been from time immemorial an article of faith that they should be broken only at the larger end, and notwithstanding that their book of faith and morals required only that all true believers should break their eggs at the "convenient end." There is no evidence in these crystal cities, however, of any rival factions corresponding to the "Big-endians" and the "Little-endians" of Lilliput. The figures are crowded together like masses of people before a hustings or at a cricket match, and the effect is passing strange. They have been formed by water which has been retained for a time in natural basins and then gradually percolated through the floor, possibly to enter into the composition of crystal cities in other sparkling caves. In this respect they resemble the basins previously described. The crystallization formed in still water, or in water which moves only downwards, passing slowly through the floor as through a dripstone, is always characterized by extreme delicacy and elaboration. The contrast between the "Lot's Wife" crypt and this is as great as the distinction between the Dead Sea and the Garden of Eden; between Gustave Doré's illustrations of Purgatory and Paradise; between Milton's L'Allegro and his Il Penseroso; between the Pink and White Terraces which until recently were the delight of New Zealand tourists, and the eruptive mask of scoria which now covers their charms.

In this remarkable cave several distinctive features are presented. The central horizontal line is well defined by an overhanging ledge, from which hang some splendid stalactites. Several of them are of extraordinarily large size, elegant form, and delicate colour. One, of pure white, on the left-hand side of the cave, rests on the head of a sturdy stalagmite which has grown from the middle of a mass of rocks, sloping down to the base. A little to the right are two twin stalactites, caricatures of the stretched out scraggy legs of some very-long-cold-and-hungry man, and the most prominent central figures of the ledge are two conical pieces tapering off to fine points, like mammoth icicles. Above this ledge the formation of lime on the dark rugged wall and roof resembles fleecy clouds in an angry sky. Below, running back into the mountain, are the Crystal Cities, fenced in with corrugated sparkling walls coped with shell-formation. A little beyond is another wall of a similar kind, also gracefully curved in obedience to natural laws, for Nature loves curves and wages perpetual war against straight lines. In the distance are five or six other mural divisions. The central one is gracefully bent like the letter S, but not quite so much rounded at the ends. Between the outer wall and the rest is an open pear-shaped space, in which are four small domes and two conspicuous figures. The principal of these is "The Queen's Statue," a pleasing stalagmitic form of pure alabaster, about 20 inches in height, and standing on a pedestal of white, shading off to brown. It does not require much imaginative power to see in this image a representation of some royal personage clad in ermine robes. The proportions are good, and the pose is exceedingly graceful. Near to the Queen's Statue is another notable stalagmitic object, formed in three sections, indicating periods of rest between. First there is a foundation of white limestone formation. Then there is a columnar growth of a few inches, with a distinct joint between it and the base. Superposed is a dome-shaped summit, not unlike the back of a human head of that kind which phrenologists call "intellectual;" and between this and the lower portion is another well-defined joint at the nape of the neck. It is peculiar, but not regal, in appearance. The topographical aspect of the Crystal Cities is something like that presented by a bird's-eye view of a piece of country, in which everything is much foreshortened, as in the case of the Katoomba Colliery and mining township in the depths of the Kanimbla valley, when seen from a cliff 1,000 feet overhead. Giant eucalypti are dwarfed to the proportions of pot plants. Tall tree ferns resemble starfish. Stalwart workmen are reduced to pigmies, and the railway seems like the double line at the foot of an account in a ledger. The various walls in the cave are supposed to encompass separate cities. The old English idea of a "city" is an incorporated place, with a cathedral and a bishop. In America all incorporated towns with a mayor and aldermen are spoken of as cities. But modern cities are not walled like those in the caves. The latter accord more with ancient cities which were intramural. For "cities" are ancient. Cain built one. Walled cities were numerous in the land of Canaan. But from the Cities of the Plain to the Apocalyptic City of Gold, with foundations garnished with all manner of precious stones, it would be difficult to imagine anything more brilliant and sparkling than the Crystal Cities of the Jenolan Caves.

Passing from the Crystal Cities to "The Show-room," about 20 yards north, several charming features present themselves for admiration, among which are conspicuous a glittering cascade, terraces of warm brown colour, reticulated; and also a pure white, delicately-made shawl hanging from the roof. The height of the crypt is about four feet, and its breadth 12 feet. The floor is mitred at the side, and between the shell borders are little forests of figures. Some elegant Doric shafts extend from the floor to the roof, which is adorned by many sparkling stalactites. The Show-room itself is a marvel of beauty. Its name indicates that it is a place of splendid exhibits, and it is appropriate. The cavern is 12 feet high, 12 feet wide, and about 40 yards long. Some of its principal features are remarkable for their elegance, and the most striking figure is distinguished by classic grace. It is a stalactite of purest white, seven or eight feet long, and from a little distance seems as smooth and round as though it had been turned in a lathe. It tapers very gradually, and its termination, which is thickened a little, rests on a stalagmite equally symmetrical, but formed in sections, each layer—of which there are about a dozen—indicating separate periods of formation and times of rest. This peculiar stalagmite stands on a dome marked by similar lights and shades. It is the centre of many varieties of cave ornamentation. For the most part the stalagmites are dumpy and poor, but there are numerous shelving platforms hung round with gorgeous stalactites, and above them are pieces of intricate formation, both floral and coral, of different classes of excellence. There are magic haunts and silent chambers with coruscations like twinkling stars on a "moon-deserted night." Nature has cast over the whole of this cave a spell of exquisite beauty.

Near to the Show-room is the cave of "The Grand Stalactites." It is only about 10 feet by 13 feet, but its massive grandeur is so impressive that the pleasure it produces is "akin to pain." Some of the stalagmites are enormous. Their summit is lost in the masses of huge stalactites which depend from the roof, and they rest on beautiful brown terraces and mounds, covered with scintillating reticulation. The elephantine stalagmites, like the stalactites which hang between and about them, and in several instances descend nearly to their base, are of dazzling whiteness. The massive pillars are so close together that the intervening spaces look like columns of jet, and thus we have "buttress and buttress alternately framed of ebon and ivory." The majesty of this cave haunts you. When the magnesium lamp is extinguished and thick darkness once more casts its veil over the magnificent scene, the vision of beauty dwells in the mind like a memory which stirs the depths of the soul.

CHAPTER XXV.

THE FAIRIES' BOWER.

A few paces only from "The Show-room" is "The Fairies' Bower," rich in grotesque lines and mystic crypts, in the purity of the formation which decorates it, and in delicacy of tints and shades. In the midst is a peculiarly-shaped stalactitic pillar resting on a dome, and which may be regarded as the Pixies' trysting-place. Then there are the "Diamond Walls," covered with millions of gems, each of which, as it flashes through the gloom, sparkles like "a rich jewel in an Ethiop's ear," or, like the very obtrusive French paste "drops" which, on her "Sunday out," light up the auricular organs of Mary Jane. It is noticeable that the walls are not of the formation which is diamantiferous. There is nothing suggestive of the diamond beds on the banks of the Ganges, the gem mines of Brazil, the rich fields of South Africa, or even the more recently-discovered diamond-bearing districts of Bingera, Mudgee, or New England, in New South Wales. There is not any gravel-drift, and the reflected light displays neither diamantine lustre nor play of colours; but, nevertheless, there is great brilliancy. It suggests, however, not so much the glories of the Koh-i-noor or the splendours of the Orloff, in the Russian Imperial sceptre, as the glitter of spangles on the dress of the acrobat.

THE SELINA CAVE.

North-west from the diamond walls is a gallery to the Selina Cave. Midway, or about 20 yards from the junction at the diamond walls, is "Cook's Grotto," named after Mr. Samuel Cook, of Marrickville. It is a lovely nook, with stalactites of rare purity and beauty. This grotto is about five feet high and from 2 to 10 feet wide. In front is a conical sloping rock with a waxy-looking reticulated surface; and the grotto itself is filled with stalactites as white as the foam of an ocean billow dashed upon the rocks of an iron-bound coast. Between two of the principal stalactites is a fine "shawl" or "scarf," made of the purest meerschaum. Over the Grotto is some pretty formation, and round about are many coloured stalactites, which make a striking contrast. To the left is a splendid alabaster stalagmite, semi-transparent, like camphor, and the little stalactites above, from which it has been formed, are of similar character, being slightly opalescent, and without any stellar reflection of light.

The Selina Cave, 30 yards north of Cook's Grotto, is about 20 feet high, 20 yards long, and 25 feet wide. It is named in honour of the wife of the Hon. E. B. Webb, M.L.C., of Bathurst, and in recognition of the interest in the caves taken by the Webb family from time to time. The walls are heavily laden with ornament. There is on the right hand side a lavish supply of variously-coloured stalactites. A large mass of formation has flowed from the fore part of the ceiling to the left, and assumed all sorts of graceful and fantastic shapes, until it reaches a shell pattern composite dwarf wall on the floor, about a foot thick at the base, and thinning off to about an inch at the top. The best of the stalactites are objects of great curiosity on account of the peculiar way in which they are embellished by small gnarled and twisted projections, and protuberances like miniature stalactitic Protea in every conceivable kind of tortuosity. In front of the cave is a magnificent stalagmite called "Lot." It is about 18 feet high and two feet in diameter, and may be supposed to represent the Patriarch after his capture by the confederate monarchs who made war against the Kings of the Cities of the Plain. It is rather large, but then it is recorded—"there were giants in those days." It is not on record why this particular column is called "Lot." Perhaps it wasthought that the briny tears shed by the son of Haran on account of the loss of his wife would be sufficient to make a second pillar of salt of similar dimensions. The roof opening into the cave is adorned with stalactites, and on the floor there are besides "Lot" two other remarkable stalagmites, which may be taken to represent members of his family. In shape they are probably quite as representative as Noachian figures of Mesdames Shem, Ham, and Japhet, that give variety to the contents of those wonderful arks which are supposed to convey to the juvenile mind the principal incidents associated with the Deluge. The roof opening into the cave is adorned with splendid stalactites, and among these is a beautiful white "shawl." The floor is composed partly of handsome basins, on the bottom and sides of which is elegant crystallization. In one corner of the roof some rich colouring sets off pure snowflakes done in lime. Among the stalactites and stalagmites are some exceedingly grotesque figures. There is also a little tablet, "Selina Cave, Feb. 7, 1881."

THE MYSTERY.

From the "Selina Cave" to "The Mystery" is about 25 yards north. In the gallery leading to this wondrous cavern may be noticed a number of small testaceous shells resembling those of cockles, embedded in the walls. The cave itself is rightly named, because of the wonder it excites by its sublimity. The spectator is first fascinated by its magnificence, and then puzzled to account for the many peculiarities and eccentricities of form which present themselves. It is an enigma. Some of the conformations appear more like the expression of vagrant fancies than the result of inanimate natural forces. Florid stalactites and floriferous rocks have become almost commonplace objects, and the visitor by this time is nearly satiated with limestone beauty; but here he finds new marvels written in mystic characters, which can be deciphered only by long and patient study. The cavern is about 12 feet high, from 8 to 20 feet wide, and about 20 yards in length. It consists of two parts. On one side is an immense mass of delicate white and rich cream-coloured formation, with numerous giant stalactites, which in purity rival virgin snow. Smaller stalactites in great profusion, are of the most perfect whiteness. But in the midst of the masses of formation are most wonderfully-contorted tubes and threads and thousands of fine lines, some like hairs and others like spun glass. There is filagree work of the most recherché kind. Some of the threads are as fine as the filaments of a spider's web, but twisted and turned in a more subtle manner. Some of the stalactites are decorated in the same mysterious way, their hirsute covering being partly pellucid and partly opaque. On the other side of the cave is a similar wonder, which still further illustrates the mysterious operations of nature, and shows with what affluence she can adorn, and how fertile are her resources.

NELLIE'S GROTTO.

About 10 yards north-west from "The Mystery" is "Nellie's Grotto," named in honour of Mrs. Carruthers, of Sydney, who visited the cave shortly after its discovery. This grotto varies from one to five feet in height, and is about

20 feet from end to end. Its beauty is bold and striking. Its chief features are taken in at a glance, and its general effect is unique. In the foreground on the right hand side is a pillar of great thickness and apparent strength, standing on a gracefully waved and rippled mound. About half of this column is stalagmitic, and was formed at different rates of speed, as may be noticed from its uneven bulk and the shape of the sections which enlarge upwards. Each succeeding portion belongs to a separate epoch commencing on a smaller circumference, until it imitates in shape its predecessor; and then the process is again repeated. The stalagmitic part of the pillar ends at about the centre of the column, and is out of the perpendicular, like the leaning tower of Pisa. From the shape of the upper section, it seems as though three or four stalactites of about equal thickness, but of various lengths, had been closely cemented together. Altogether this pillar is a very bold and beautiful object. At the other end of the grotto are two columns composed in a somewhat similar manner, but of much smaller dimensions. They also rest on blocks of formation, and are surrounded by a number of splendid stalactites, some of which descend half way and others to the base. Between these two extremities, there are on the floor about a dozen little cones of white matter like sugar-loaves, but not so smooth. On four of them rest the points of magnificent tapering stalactites, straight from the roof of the grotto, and in a line with these and the two extreme columns are stalactites of various lengths, some plain and some richly ornamented. The interior of the grotto is also enriched with similar kinds of beauty. A little distance away to the left is to be seen on another bold rocky ledge a second series of small stalactites, suggestive of the commencement of another grotto similar to that which now wins so much admiration. There is also a very remarkable stalagmitic formation which resembles a wax candle burning before a shrine of Purity. It is with reluctance that the tourist turns from this lovely grotto and wends his way towards "The Vestry." He feels as though he could hardly refrain from taking a last fond look, even though he should run the risk of being converted into a limestone pillar.

CHAPTER XXVI.

THE VESTRY, THE JEWEL CASKET, THE BRIDAL VEIL, AND THE FLOWERING COLUMN.

About 25 yards north of Nellie's Grotto is "The Vestry," a cavern about 12 feet high, 8 feet wide, and 20 feet in length. It is called The Vestry, because of its propinquity to some very fine cathedral-like architecture, and not because any minister requires it to robe himself in, or because it is used for the meetings of any parochial assembly. This Vestry runs east and west, and there is here a large area of unexplored caves. Indeed, it may be said that there are in almost every part of the mountain openings which indicate the possibility of the existence of new and lovely caverns. It is also probable that in process of time it will be found that all the principal chambers are connected by passages which by a little enlargement may make intercavernous communication complete between all the best known caves. That this is likely may be gathered from an incident which occurred a few months ago. There were then at the Cave House four black cats, slightly marked with white. On being petted, the playful animals would not only purr their thanks, but also follow like dogs. One day the most venturesome of them went with a party into the Lucas Cave, and travelled with them a considerable distance before it was missed, and then it was thought that it had returned to the daylight. In the evening, however, it was not in its accustomed place on the hearthrug, and about the middle of the following day it was discovered in the Imperial Cave on the other side of the Grand Arch! The attention of the party was attracted by the doleful cries of the animal, which had by that time realised the fact that it had lost itself in a dangerous place. Some doubt might have existed as to the identity of the cat found in the Imperial with that lost in the Lucas Cave, had it not been that with the party that took it in were two boys, who had carried it now and then, and permitted its extremities once or twice to come in contact with their lighted tapers. They were able to identify it by "the drips of sperm on its back, its burnt whisker, and the singe on the tip of its tail!" Thus a very interesting fact was established which but for the accidental brandings would have been doubtful. The visitor returns to the Fairies' Bower and the Grand Stalactites junction; and about 20 yards north-west from the junction he arrives at the Crystal Palace, which is fenced in by wire-netting in order to protect its marvellous grandeur from that class of sightseers who appear to be unable to look at anything except they can place their unwashed paws upon it—in which case impressions are mutual.

There are three distinct types of beauty in this Palace—the simple, the compound, and the elaborately complex. The simple forms are extremely massive. To call them "pretty" would be to convey an entirely erroneous idea of their quality. The stalagmitic features are immense. One of them is a gigantic pillar, built up in sections from an enormous basement on a rocky mound, over which it has flowed like milk-white lava. About a foot or so above the crest of the rock it is several feet thick, and at one time its upper surface formed a disc or table. On this flat top was gradually formed another white mass of somewhat smaller proportions, tapering towards the top, or second platform, from which rises another stalagmitic section, expanding from the base, and this process is repeated to the very summit, so that the beautiful white pillar has a serrated appearance. For unsullied whiteness and peculiarity of structure it will bear favourable comparison with the most remarkable pillars in the cave. Near to it is another pyramidal-shaped mass of even greater bulk, which tapers as it rises towards the stalactitic formation, and harmonises with it in grandeur.

From the right hand side of the base another and smaller stalagmite rears its head, and immediately above it, and to the right of it, are masses of formation hanging like stalactites of various lengths, and bound together in solid but graceful combination. Farther still to the right is an immense stalactite, shaped like the body of a kingfish minus its tail. The floor of this part of the Palace is also very attractive, by reason of its graceful curves and undulations, and miniature rippled terraces. The more complex part of the Palace to the left begins with a magnificent piece of shawl formation, from the lower end of which depends a splendid stalactite. The shawl is draped at an angle of about 22½ degrees, and in a line with its principal portion the wall is hung with marvellous stalactites, one of which is of great length, and clear as crystal. Then there is a cavernous place, from the shades of which emerge rounded masses of white formation, fringed with myriads of stalactites. From behind these the same kind of ornamentation is repeated again and again until a ledge of rocks is reached, which slopes down to a marvellously beautiful stalagmite several feet high, and which rises from the floor immediately underneath the stalactite at the end of the shawl before mentioned.

The general impression left by this part of the Palace is that its grandeur is different from that which distinguishes other portions of the caves, but it would be as difficult to say in what the peculiarity consists as it would be to describe the general appearance presented by different turns of a kaleidoscope. The most intricate part of the Palace is distinctive enough to leave a separate memory. The wall is covered with masses of brilliantly white formation, with stalactites all about them, some short, some long, some tapering like icicles, some straight like pipe-stems; most of them pellucid, and some like iridescent glass. Some of the "shawls" are delicately tinted, and present a charming appearance. There are deep brown and delicate fawn-coloured banks, which seem as though they were covered with a stony network. Little caves at the sides are partly filled with drifts of glistening snow. Some of the ledges are covered with white stucco, with delicate fringes. Many of the stalactites are charged with water, and the drops coquette with the light and rival the glitter on the walls. There are stately and elegant shafts of alabaster from floor to ceiling, coloured stalactites and stalagmites nearly meeting. At every glance the eye is pleased with new and curious forms and rich combinations of colour. Masses of the formation are fringed with contorted threads and pipes, and on the foreground are some curiously-shaped masses like snow, with delicate frost work and projections like frosted hairs all over their surface. These are for the most part opaque, but the predominant features are crystal.

The distance from the Crystal Palace to the Jewel Casket is about 15 yards north-west, through a hall about 25 feet high and about 15 feet wide. The Casket itself is a horizontal fissure in the rocks, about 8 feet by 12, filled with brilliants of various hues. Its splendour is enchanting. Overhanging stalactites guard the Casket, and form as it were bars of alabaster, opal and crystal, and through the spaces may be seen many different varieties of crystallization. The floor is carpeted with jewels, set off with sparkling masses like frozen snow. Some of the gems are white like diamonds, some coloured like cairngorms, and other varieties of rock crystal of even more delicate tints, and numbers of them are clear and translucent. Some of the ornamentation is of a rich brown. The impression produced is that nothing could possibly be more brilliant and entrancing than this rich casket; and yet, remembering how many times previously he has come to the same conclusion and subsequently found he had miscalculated the magnificence still in reserve, the visitor hesitates to accept the Jewel Casket as the ne plus ultra of cave magnificence. Then there is the Bridal Veil, about 10 feet by 2 feet—a wonderful piece of delicate tracery imitating fine lace—not écru, but white as the fairest emblem of a blameless life. Here are numerous terraces in deep brown and fawn colour covered with spangles which glitter like broken-up moonbeams on the wavelets of a summer sea or the phosphorescence which, in the wake of a ship, mocks the stars. The Flowering Column comes next—a huge mass of formation 25 feet high, branching off into all sorts of shapes graceful and grotesque. It is about eight feet wide in the centre, of a rich brown colour shading off to a brighter and lighter hue. This pillar is covered with remarkable little figures like flowers natural and fanciful, and near to it is a series of imitation cascades in regular sequence which simulate so much natural force that they might be taken as an illustration in lime of "how the waters come down at Lodore." These cataracts or waterfalls are now for the most part dry; but at one time the supply of liquid or semi-liquid limestone, of which they are formed, must have been very abundant.

CHAPTER XXVII.

HOW CAVES ARE MADE—THE WORK OF AGES.

There remain now to be described but four of the caves ordinarily frequented by visitors. These are "The Garden Palace," "The Stalagmite Cave," "The Gem of the West," and "The Fairies' Retreat," with "The Queen's Diamonds." But, as already intimated, it is impossible to foretell what visions of loveliness may be disclosed by future explorations. Quite recently the curator has wormed his way into another splendid cave of large dimensions and great beauty. An opening of about eight feet leads to a small chamber 14 feet wide and from four to eight feet high. The floor is pure and sparkling. There are some very pretty stalactites and pieces of formation hanging from the roof, with transparent pipes and straws terminating in little hair-like projections. From this cave there is a passage 10 inches by 14 inches high and then 14 inches high by 10 inches wide, extending about 21 feet, and opening on to a room with a floor of velvet-like coral. As the foot rests on it the sensation resembles that produced by walking on a new Brussels carpet, or stepping on a frosted lawn of buffalo grass, which slightly crunches beneath the feet.

This cavern is about 25 feet by 35 feet, and from 2 feet to 10 feet high. From it there is a fall of about 20 feet, opening into another chamber, to the left of which is a pretty fimbriated tray, 10 feet by 12 feet, filled with little knobs of formation, with points so sharp that any pressure of the hand would cause pain. These nodes and points are almost as clear and spotless as the drops of a glass chandelierbefore they have been converted into fly-walks. Then there is another tray or basin with looped edges containing crystals which are a little "off colour;" and yet another, with escalloped border and formation of a milky hue—that is, pure milky—opaque white, not cerulean blue. This basin is about 14 feet by 16 feet. The roof immediately over the basin is like delicate coral. To the right is a sort of illusion representing solidified water round about black limestone pebbles. This crystal has been formed by a run of water from a rock, the summit of which is about 20 feet away, and which slopes down from the wall at an angle of about 45 degrees.

The sloping rock is beautifully reticulated and marked by curved lines, which gradually diminish towards the base, where the formation changes to nodules and curiously-formed, irregular combinations of the preceding figures massed together. From the ceiling descend various formations of cave decoration, some clear as the skin of a Mayfair beauty at her first ball, others of a delicate fawn colour, and the remainder tinged with oxide of iron. There is a range from the delicate lily of the valley to the rough bronze of the muscular navvy, not to speak of the dusky brown of the unwashed sundowner who arrives at a "station" when "the shades of night are falling fast," and whose motto is not "Excelsior."

The next chamber is about 14 feet by 12 feet, at the end of a slight declivity as white as snow. It is in the form of an ordinary retort, and is succeeded by another chamber of somewhat similar contour, the principal ornamentation being on the floor. Thence the course is upwards, and the most attractive formation is from the roof. The stalactites are in clusters, and for the most part small. Some of them are like twigs, but clear as a limpid stream. Here also are stalagmites about nine inches high, formed on the ends of huge rocks. In addition to these are other smaller stalagmites near to a hole about 14 feet deep, and on the floor are fossil remains. In this hole are passagesunexplored. Hard by is a pretty "shawl" hanging from the roof, and beyond it a rock about three feet high, the edge of which is covered with a fine substance like down, which, when blown upon, flies about like the winged seeds of thistles. Descending from this cavern, about 10 yards through a narrow passage, there is a steep fall of nearly 10 feet, which leads to a chamber, the roof of which is about 10 feet high. On the right hand side of this cave are some very white shawl pattern formations and stalactites, and to the left are fine shawls, clear as glass, from four to eight feet long, and from 6 to 12 inches deep. Just beyond is a crisp, velvety floor, like that which characterises a chamber previously described, but not of the same colour. This floor is in some parts very red, as though it had been coloured with clay, and other parts are like glass stained with red ochre. The stalactites are thin, and formed in all sorts of peculiar shapes. The floor is uncommon, being of a piebald character, appearing as though buckets of solution of lime had been cast upon it, in the midst of some large stalactites that give character to the cave, which is about 40 feet by 40 feet. Passing on you come to a pretty chamber ornamented with stalactites, composed mostly of yellow crystal. The cave then runs nobody at present knows where. There are unexplored chambers all around. This cave is named after the Government geologist, and is to be known as the Wilkinson Cave.

Mr. C. S. Wilkinson has contributed some valuable matter to Cave literature. His account of the Jenolan Caves, written for the Government, contains several beautiful passages bearing upon the physical and chemical agencies at work in the formation of limestone caverns. And here it may be interesting to refer to one or two other authorities also on the same subject. Dr. Wright, an American scientist, says, "There can be no doubt that the solvent action of water holding carbonic acid in solution" is the primary agency concerned in the formation of limestone caves. "Limestone," he says, "is not soluble in water until it combines with an additional proportion of carbonic acid, by which it is transformed into the bicarbonate of lime. In this way the process of excavation is conducted until communication is established with running water, by which the mechanical agency of that fluid is made to assist the chemical. Little niches and recesses, which seem to have been chiselled out and polished by artificial means, were formed in this manner; for when these points are strictly examined, a crevice will be observed at the top or at the back of them, through which water issued at the time of their formation, but which has been partially closed by crystals of carbonate of lime or gypsum." Dr. Wright, referring to the different conditions and different periods of cave formation, says:—"The sulphate of lime, which is known under the name of gypsum, plaster of Paris, selenite, alabaster, etc., exerts a much greater influence in disintegrating rock than the sulphate of soda. The avenues in which gypsum occurs are perfectly dry, differing in this respect from those that contain stalactites. When rosettes of alabaster are formed in the same avenue with stalactites, the water which formed the latter has for ages ceased to flow, or they are situated far apart, as the former cannot form in a damp atmosphere."

Mr. Wilkinson also alludes to separate periods of formation in connection with the Jenolan Caves. He says:—"There appear to have been two distinct periods during which stalactitic growth formed; one of comparatively remote age, and very local in character, being chiefly confined to the caves known as the Lurline and Bone Caves; and another but recent and still in operation. The older growth is essentially of a stalactitic type, and the stalactites are remarkably thick, though in one or two cases a huge stalagmite is to be seen. The newer growth exhibits every fantastic and beautiful form known, from the thin hollow reed and transparent veil to the snow-white dome stalagmites, the crystal-fringed pool, the wave-lined floor, and the crooked-fringed shapes that are turned in all directions." But there is one passage in Mr. Wilkinson's account which takes us far beyond the time when the limestone mountains were formed, and describes a complete circle of natural transmutation and reproduction, and which may be appropriately quoted in connection with the cave which bears his name. Here it is:—"First, the decaying vegetation of some ancient forest is invisibly distilling the gas known as carbonic acid; then a storm of rain falls, clearing the air of the noxious gas, and distributing a thousand streamlets of acid water over the surrounding country, and which, as it drains off, not only wears the rocks it passes over, but dissolves them in minute quantities, especially such as contain much lime, and then, laden with its various compounds, flows off to the distant sea, where reef corals, lying in fringing banks round the coast, are slowly absorbing the lime from the water around them, and building the fragile coatings that protect them during life. Slowly as the land sinks the coral bank increases in height, for reef corals can only live near the surface of the water; and soon a considerable thickness has been obtained; while below the upper zone of live corals lies a vast charnel-house of dead coral coverings; then comes a change; suitable temperature, or some other essential condition, fails, killing out all the corals, and through long ages other deposits accumulate over them, gradually crushing and consolidating the coral bank into a firm rock. At last a convulsion of the earth's crust brings it up from the buried depth in which it lies, leaving it tilted on its edge, but still, perhaps, below the surface of the ground; rain, frost, and snow slowly remove what covers it, until it lies exposed again to the sunlight, but so changed that but for the silent but irresistible testimony of the fossil forms of which it is composed, it were hard to believe that this narrow band of hard grey rock was once the huge but fragile coral bank glistening in the bright waters with a thousand hues. And now the process is repeated; the decaying vegetation of the surrounding forest produces the carbonic acid, the rains spread it over the ground, which is now the most favourable for being dissolved, and the consequence is that the acid water saturates itself with the limestone rock, and whenever the least evaporation takes place, has to deposit some of its dissolved carbonate of lime in one of the many stalactitic forms, before it can flow off to the sea and distribute its remaining contents to fresh coral banks. Thus the old coral reef melts away far inland, and the lime that formed the coatings of its corals is again utilised for the same purpose. What a simple succession of causes and effects, and yet before the circle is completed long ages of time have come and gone; and what a fine example of the balance between the waste and reproduction that takes place in Nature!" And thus the diurnal motion of the earth and its annual journey round the solar circle, as well as the repetitions of history, have impressive geological analogies. How many hearts have begun to beat—how many have throbbed with passion and ambition, and waxed cold as an extinct volcano in the years required to form a small stalactite? How many ages have come and gone since the Jenolan Caves were coral reefs in the azure sea?

CHAPTER XXVIII.

THE GARDEN PALACE—THE STALAGMITE CAVE AND THE GEM OF THE WEST.

The "Garden Palace," about 14 yards north from the Flowering Column, is remarkable for the beauty of its proportions and the charming grace of its arches and dome. It has on the left hand side a magnificent stalactite descending from the roof, and coming to a fine point on the top of a stalagmite, which rises a short distance from the floor. There are also many other stalactites and stalagmites of rare proportions. Near to it is a fascinating little crypt that can be peered into from a small aperture in the wall of the passage, which is here about 25 feet wide and 14 feet high. This part of the "Palace" is about 4 feet by 4 feet 6 inches and about 10 inches high. There is no association about this portion to vividly recall the elegant building which adorned the Inner Domain, or even the grandeur of its ruins, which the fierce flames could not consume. It is simply a charming little peepshow filled with the most dainty specimens of crystallization, the purest stalactites, and the most elaborate decoration. In front is a stalagmite called the "Prince's Statue." Most of the stalactites are transparent. The stalagmites are white as snow, and some of them sparkle with an external coating like hoar-frost. There are small globular pieces covered with tubular spikes, like those of the echinus, but as fine as the stings of bees or the antennæ of butterflies. In the centre is a little colour of a roseate hue, and the most prevalent forms resemble transparent flowers and plants which rival the skill of the glass-blower, and surpass the most delicate work of the artificer in gold and silver. This crypt is like a dreamland treasure-house filled with spoils of art and fancy.

The Stalagmite Cave and the "Gem of the West" are about 30 yards north of the Garden Palace, through a hall about 12 feet high and 14 feet wide. The stalagmites are magnificent, and the walls are adorned with glittering formation of delicate tracery. There are some fine specimens of stalactites, and the distance between two of them has been carefully measured with a view to accurate observation as regards the rate of future growth. Not far from these are some remarkable stalagmites, formed on and about a sugarloaf-shaped mound. The uppermost one, which appears incomplete—not having yet received its apex—is composed of five sections. Another close by is formed of eight or nine sections, and one lower down of about the same number of distinct portions. Like some other stalagmites already described, their individual sections represent separate periods of formation and of rest.

One of the most beautiful of the large stalagmites in the caves is the Alabaster Column.

The "Gem of the West" is in every respect worthy of its name. It is one of the most attractive caves in the series, and calls forth expressions of delight the moment it is illuminated by a magnesium lamp. It is carefully protected by wire netting, and retains all its pristine loveliness. It occupies but a small space, being seven or eight feet high by about five feet wide. There is a considerable amount of formation on the roof, extending down the wall to the ledge of a rock, the flat under-surface of which forms the roof of the cave. From it descend numbers of stalactites of various lengths, as clear as crystal. In between these are numerous small glass-like ornaments, and here and there are little rifts filled with a substance like drifted snow and sleet. The stalactites are not all clear; some of them resemble alabaster, and their shapes are very grotesque. Many are straight as reeds; some are bulbous, and several are combinations of straight pipe and bulb. A few have grown obliquely, but many of the smaller ones are contorted in the most extraordinary manner. On the other side is a formation like a miniature Niagara, with "wild shapes for many a strange comparison," and forms of exquisite beauty.

"Full many a gem of purest ray sereneThe dark, unfathomed caves of ocean bear."

But ocean caves do not contain anything more pure and captivating than the splendours of the Gem of the West.

CHAPTER XXIX.

THE FAIRIES' RETREAT.

It is about 40 yards north-east from the Gem of the West to the Fairies' Retreat. The passage is from 8 to 20 feet high and from 6 to 14 feet wide, and rather damp. The visitor ascends a short ladder to a rocky ledge, where there is a small opening and a narrow passage, along which he has to wriggle his way in a recumbent posture and with his feet foremost. Long before he has reached the immediate entrance to the Retreat he begins to think Puck has led him a "pretty dance," and he has gained some idea of the least pleasant sensations incident to cave exploration. When he has completed the journey in a doubled-up posture, and is placed in an attitude scarcely less uncomfortable, curved like a boomerang, he feels as though he would give the world to be able to stretch himself. But a slight pressure upwards reminds him of the superincumbent mountain, and so he feels like a prisoner with billions of tons above him and the rocky base below. He begins to grow hot, and would give anything to be in a place capacious enough to enable him to expand and breathe freely. However, the lamp is turned on, and for a moment or two he is lost in admiration of the scene. He might fancy himself Sindbad in the Diamond Valley, or think that the cave-keeper possessed the lamp of Aladdin, or that he had come upon enchanted land. This Retreat extends S.E. about 20 yards. Its entrance is about 2 feet by 20 inches at the embouchure, and it widens a little towards the end. It is about four feet wide and three feet high, and is filled with glittering cave gems and alabaster flowers, and myriads of figures which sparkle with brilliants. But what are the brightest jewels and the choicest flowers to ease of body and mental serenity? Many of the fair sex have visited this Retreat, carefully tutored and assisted by the curator. It may be appropriately and pleasantly inspected by agile sylphs and dapper little men who affect a contempt for muscular development and insist upon being gauged by Dr. Watts's standard, but ladies who are massive and gentlemen who are portly and plethoric will, when making their exit, caterpillar fashion, think it very absurd that so splendid a spectacle should have so mean and inconvenient an approach.

THE QUEEN'S DIAMONDS.

After seeing the Fairies' Retreat there remains but one other surprise, and that is "The Queen's Diamonds." These are in a casket easily accessible, and the opening to which is about three feet wide by one foot high. The jewel case itself is about four feet wide, three feet deep, and 12 feet long. When the light is turned into it the brilliancy of the scene is perfectly dazzling. The prismatic formations are wonderful, and the blaze of magnificence mocks the descriptive power of either pen or pencil. It is "labyrinth of light" which appeals to the imagination with rare force. Edgar A. Poe worked up an excellent sensation in his story of "The Gold Bug." The way in which Mr. William Legrand became possessed of the scarabæus with scales of bright metallic lustre, and of the scrap of paper which contained mysterious directions leading to hidden piratical plunder by Kidd, is not more interesting to the general reader than cryptography is to the student. The enthusiastic way in which the curator speaks of this cave and its distance and measurements recalls to memory the exciting incidents connected with the death's-head, the gold bug dropped through its eye-socket, the taping of the distance from the fall of the scarabæus to the hidden wealth, the hurried digging, and the discovery of the buried treasure. "As the rays of the lantern fell within the pit, there flashed upwards a glow and a glare from a confused heap of gold and of jewels that absolutely dazzled our eyes." The feeling produced in that case was exhaustion from excitement; but the sensation caused by a glance at the brilliance of the Queen's Diamonds is one of intense gratification. It is a most vivid and lustrous spectacle. The crystals are in clusters grouped together like the petals of flowers, and these flower-shaped forms combine with others of a similar kind, and constitute elaborate floral masses. They are much more difficult to decipher than was the cryptography left by the pirate Kidd. As Brewster puts it, "though the examination of these bodies has been pretty well pursued, we can form at present no adequate idea of the complex and beautiful organization of these apparently simple bodies" Of the 1,500 or more different crystals known to science, nearly half are composed of carbonate of lime, but "The Queen's Diamonds" are certainly among the rarest. The crystallized forms in the caves are very numerous. Some of them are irregular, on account of the substance not having been sufficiently divided before its deposition, or because of inadequate space or insufficient repose, but for the most part they are regular and perfect of their kind. None, however, are more regularly formed or more pronounced than "The Queen's Diamonds." It would be difficult to describe their geometric shape. The separate fragments of each cluster vary from about an inch to a fraction thereof. They are like three-sided prisms, tapering to points at the ends. The edges are sharp as knives from the centre to the upper point; but from the centre to the end which is joined to others, the sharp edge is replaced by a smooth surface, as though a cut had been made with a razor. The upper ends of these prisms are clear as glass; the lower ends are a little cloudy. The brilliancy of the combination is marvellous.

CHAPTER XXX.

GENERAL IMPRESSIONS.

When the Queen of Sheba heard of the fame of Solomon she went to Jerusalem with a great train, with camels that bore spices and large quantities of gold and precious stones, and fully satisfied the curiosity commonly attributed to her sex. She proved the wisdom of the far-famed monarch, admired his house and its appointments, the apparel of his servants, the attendance of his ministers, and the magnificence of his daily table; and, according to Josephus, she said: "As for the report, it only attempted to persuade our hearing, but did not so make known the dignity of the things themselves as does the sight of them, and being present among them. I, indeed, who did not believe what was reported by reason of the multitude and grandeur of the things I inquired about, do see them to be much more numerous than they were reported to be." Or, as the verdict of "the Queen of the South" is given in the Authorized Version of the Old Testament, "I believed not the words until I came, and mine eyes had seen it; and, behold, the half was not told me." A similar testimony will be borne by most visitors in regard to the magnificence of the Jenolan Caves, and "the multitude and grandeur" of the objects which excite surprise or challenge admiration at almost every step. Jenolan is a veritable wonderland, as well as a most interesting geological study. It presents features sufficiently sublime to touch the deepest chords of the human heart: forms sufficiently graceful to charm the artist; situations affording novel material for the romancist; configurations, transmutations, and fascinations to move the soul of the poet; and vast, silent cathedrals which inspire a feeling of devotion, for—

"Nature, with folded hands, seems thereKneeling at her evening prayer."

Some of the best photographs which have been taken of the most prominent features of the caves give an idea of their grandeur, but it is only a poor one. They convey no impression of their delicate sheen and dazzling beauty, of the gradations of tint and colour; of the mystic crypts and charming contours. A good photograph may aid those who have seen the caves to fill in details, but even to the most imaginative person who has not had the pleasure of a personal inspection it cannot possibly convey anything like an adequate sense of the thousand and one charms which elude both the photographer and the artist. There is as much difference between the pictorial illustration and the reality as there is between a vacant stare and an eye filled with lovelight and sparkling recognition. And the same remark will apply in greater or lesser degree to verbal description. Words are altogether too poor, and it may be folly to attempt to describe the indescribably beautiful. Among the numerous inscriptions in the cave book, a visitor has given his judgment upon this point very bluntly. He says, without periphrasis or euphemism—"The man who would attempt to describe these caves is a fool." Still, these articles as they appeared in the Sydney Morning Herald, have at least brought the caves prominently before the public, and perchance in their present form may be of use to future visitors.

On leaving the Cimmerian gloom of the Imperial Cave, and emerging into the clear daylight, the sensation is strange, for after being two or three hours entombed in the heart of the limestone mountains, the darkness seems as natural as night. It is joyous, however, to be back again in the sunshine, and to find that—

"There is a tongue in every leaf,A voice in every rill."

CAVES UNEXPLORED.

And as you glance once more along the limestone mountain ridge you wonder what hidden beauties yet remain to be revealed. To the north from the Devil's Coach House numerous caves are known to exist, and it is probable that some of them may present features more remarkable than any yet discovered. The creek, which runs quietly along, has on its way some oblique outlets before it sinks into the earth, and recalls, with its surroundings, the pleasure-place of Kubla Khan,—

"Where Alf the sacred river ranThrough caverns measureless to manDown to the sunless sea."

The first of these caves is very deep, with a steep ascent. The curator has penetrated it to a depth of 160 feet. He was lowered into one shaft 100 feet perpendicular, and found in it a number of interesting bones, which he sent to the museum of the Department of Mines in Sydney. One of them resembled the tusk of a tiger, and was thought to be too large to have been in the jaw of any Australian animal extant. The next known cave is called the "Glass Cave," on account of the transparent beauty of its adornments, which are equal to those of the Imperial Cave. The third is unexplored, but there is reason to believe that it is very extensive. The fourth, which seems to run southward, is also unexplored. Some time ago the curator was lowered into it, a depth of 80 feet, but he has not been able to make any examination of its interior. Next is the Mammoth Cave, so called because of its vast chambers. One of these is estimated to be upwards of 300 feet high, 100 feet long, and 100 feet wide. It contains a large amount of formation, the prettiest portions of which are about 200 feet from the floor. The roof is so high that the magnesium lamp is hardly strong enough to bring it into view. There is a very long and wide chamber leading from this towards the south, with a large number of "drops" of from 100 to 150 feet, many of them unexplored. The length of this chamber is about 10 chains—that is, one-eighth of a mile. From this the curator was lowered into another chamber of vast proportions, and from 60 to 100 feet deeper down, through solid limestone. At the end of its undulating floor he came to a river about six feet wide and nine inches deep, the water of which was running in a strong stream. Round about are many little caverns full of bones. The next is the Bow Cave, to which reference has previously been made. It is a small cavern, and, as has already been explained, there were found in it six bullock bows, together with two harrow pins, and a pair of hinges; supposed to have been "planted" there by McEwan, the bush-ranger, about the year 1839. This cave has about it numerous small drives not yet explored. Farther on is a pretty cave, running, with a gentle slope, two or three hundred yards into the mountains, and containing numerous chambers and water-holes. Five or six of these chambers only have been explored, and they contain some very handsome stalactites. Farther on are two or three other caves, to the entrance of which only the curator has been. Above all these caves a strong stream of water sinks suddenly into the ground. It is believed that this feeds the river in the Mammoth Cave, and afterwards flows into the Imperial, and comes to the surface again in Camp Creek, on the other side of the limestone range, where it bursts up suddenly from between the boulders, as though there were underneath them a broken 42-inch city water-pipe.

On the south side of the Grand Arch are several known caves. The curator has been lowered about 100 feet into one of small dimensions, with a large chamber from it, containing many sonorous stalactites of large proportions, as well as much elaborate formation. Next to this is a cavern called the Specimen Cave. It is about 20 feet deep, with a large number of bones on the floor, which has been broken away. The fracture shows that it was largely composed of red clay. The broken face is full of bones. This cave would be worth special examination, because it is believed many of the bones belonged to animals which have disappeared from New South Wales. Farther on is a cave into which, five years ago, the curator was lowered about 240 feet. At a still greater distance, three miles from the Cave House, is a cave, with a strong stream of water flowing out of it. That is the farthest limestone visible on this side of the mountain. The stream sinks into the ground at a short distance after it leaves the mouth of the cave. It is believed that this water runs underneath all the caves on the south side into the Lucas Cave, and is not seen again until it re-appears under "The Bridge." The range of limestone rocks seen from the Cave House extends north and south five-and-a-half or six miles. At each end water in considerable volume sinks into the earth suddenly and re-appears in caves near the centre of the valley; the northern stream flowing through the Imperial Cave, and the southern through the Lucas Cave. Both streams come to the surface in Camp Creek, and chatter away towards the sea.

CHAPTER XXXI.

CONCLUSION.

The journey from Sydney to the Caves is long and expensive, but the route is interesting all the way. Thirty-six miles from the city after crossing the Emu Plains and the magnificent bridge over the Nepean at Penrith, 87 feet above sea-level, the train begins to climb the mountain range, and after travelling 52 miles it attains an altitude of 3,658 feet. The first zigzag up Lapstone Hill brings into view a splendid panorama, and, notwithstanding the disregard of railway surveyors for fine scenery when it is placed in the balance against economical construction, there are many glorious glimpses to be obtained from the carriage windows during the journey between Sydney and Tarana. If the traveller has time to stop en route he will find much to interest him round about Lawson and Katoomba, Blackheath and Mount Victoria. The great Zigzag into the Lithgow valley is one of the most remarkable feats of engineering in the world. There is a fall of about 230 yards in five miles of running. Even after this sudden descent the country is still very elevated, and at Tarana, where it is necessary to take coach for the Caves, it is upwards of 2,500 feet above the level of the sea.

The train which leaves Sydney at 9 o'clock in the morning arrives at Tarana at 4.15 in the afternoon. By coach or buggy from Tarana the little agricultural township of Oberon—distant about 18 miles—can be approached comfortably the same evening. At Oberon there is awell-kept hotel, which reminds one of the best village hostelries in the old country. Here it is customary to stay the night. Bidding good morning to your host after a moderately early breakfast, you can drive to the door of the Cave House just in time for lunch.

The road from Tarana to Oberon is well made and metalled, and with a pair of good horses you can travel at a spanking pace. From Oberon to the Caves the road is also good, but not so wide as that from Tarana to Oberon. A considerable portion of it may be described in fact as an excellently kept bush track. The road down the Zigzag is, as already mentioned, a trial to the nerves of timid people. It is much too narrow, and ought to be widened by cutting still farther into the mountain side, building up the retaining wall more substantially, and paying greater attention to drainage. A cable tramway would then make the transit easy and pleasant.

A light railway to Oberon would probably give as good a return as nine-tenths of the mileage on our railways, and if the tramway from there to the Caves would not pay immediately it would ultimately create settlement and traffic, and in the meantime be an important factor in increasing the traffic on the 140 miles of railway leading to it from Sydney.

Until a short time ago the Caves were completely cut off from rapid communication with the outer world, but now they are in telephonic communication with the telegraph system of the colony.